TEACHER'S GUIDE

SEA SOUP

Discovering the Watery World
of Phytoplankton and Zooplankton

Betsy T. Stevens

Illustrated by Rosemary Giebfried

Tilbury House, Publishers
Gardiner, Maine

To Joel and our wonderful, blended family.

Tilbury House, Publishers
2 Mechanic Street #3
Gardiner, Maine 04345

First edition, October 1999.

10 9 8 7 6 5 4 3 2 1

Design and layout: Rosemary Giebfried, Bangor, Maine
Illustration: Rosemary Giebfried, Bangor, Maine
Editorial and production: Jennifer Elliott and Barbara Diamond
Printing and binding: InterCity Press, Rockland, Massachusetts

Table of Contents

Acknowledgments

My interest in "drifters" began as a child growing up on the Connecticut coast. I remember sailing at night and asking my dad, "What makes the water glow?" During college my major professor, whale researcher Roger Payne, encouraged my interest in the marine world. After decades of teaching about aquatic life and hours of watching the eerie glow in the bow waves of tall ships, I now have had a chance to write this teacher's guide and share my enthusiasm for tiny drifters. Thank you to my family and colleagues along the way for their support in all my educational endeavors.

Creative ideas come from inner sources that were shaped by years of experience. To those writers and teachers whose work inspired my creativity, thank you.

To Mary Cerullo, Jeanne Meggison, and my daughter, Amy Bayha (all inspiring, experienced educators), thank you for your patient review of contents and context of this teacher's guide.

Thanks to Alan Lishness at the Gulf of Maine Aquarium for his support on this project and his permission to use an activity from the aquarium's curriculum, *Space Available: Aquatic Applications of Satellite Imagery.*

Jennifer Elliott of Tilbury House supplied large doses of patience and support, while Bill Curtsinger and Mary Cerullo's enthusiasm kept me going. Rosemary Giebfried's fine illustrations and graphic layout were the frosting on the cake. Thank you.

To Joel for his unconditional love and to whom this project is tangible evidence that he was correct: I have failed retirement. Thank you.

SEA
SOUP

Introduction

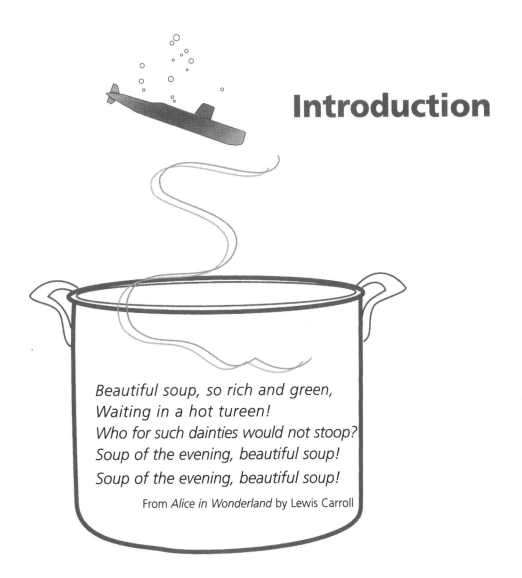

Beautiful soup, so rich and green,
Waiting in a hot tureen!
Who for such dainties would not stoop?
Soup of the evening, beautiful soup!
Soup of the evening, beautiful soup!

From *Alice in Wonderland* by Lewis Carroll

As we travel in our miniaturized submarine down into the sea soup, a glance upward through the water's surface reveals seabirds skimming deftly above the watery realm, their wing tips almost touching the waves. Sharp eyesight and quick reactions allow the birds to snatch small prey fish from just beneath the surface. Looking through our portholes we see a huge Goliath, the giant of the sea, a baleen whale cruising the banks devouring tiny fish and shrimp by the tons. A dark shadow moves overhead. A commercial fishing vessel pays out its bottom trawl to catch groundfish such as haddock and cod, flounder and hake. We need to move quickly to avoid being trapped in the net.

But the going is slow. The water is thick with tiny organisms. What are they? How do they survive? Are they important?

An Introduction to the Drifters

Of all the living tissue produced in the open oceans and seas, more than 90 percent is plankton! What are they? The word *plankton* is derived from the Greek word "planktos" which means "drifter." Plankton includes any *aquatic organism* living unattached and lacking sufficient swimming power to resist most water currents. More simply said, plankton are plants (phytoplankton) and animals (zooplankton) that float at the mercy of the currents or have limited swimming abilities. Many *plankters* are microscopic and single-celled; others may be half a meter (19 inches) or more across. Some look like miniature versions of well-known creatures such as shrimp. Others look like tiny alien craft from another planet or minuscule ornate pillboxes.

The Importance of Plankton

Phytoplankton are the most significant photosynthesizers in the marine ecosystem. The tiny, single-celled organisms, like the grasses on land, absorb sunlight and turn the sun's energy into organic material through *photosynthesis*. In the process they create a rich food supply for *zooplankton* and other plant-eating animals and, as a byproduct, they produce oxygen. This critical process of converting sunlight into organic material is called primary production. In the ocean attached plants only exist where sunlight penetrates the water. Under the best of conditions no attached plants can live and photosynthesize at ocean depths below 20 meters (65 feet).

Zooplankton are not just wanderers; they don't always follow where the currents lead. Scientists have discovered that at least some of these drifters spend more than 80 percent of their time moving under their own power. Every night, when predators can't see them, zooplankters swim up towards the water's surface to feed on phytoplankton. As the sun rises, they sink into deeper water, sometimes traveling up to 300 feet each way. Using sidescan sonar, marine scientists are able to track the movement of zooplankton from the ocean depths to the surface and back.

All fish and shellfish are ultimately dependent upon phytoplankton, but zooplankton are the intermediary category of marine life that serves to transfer energy captured by phytoplankton to other marine animals.

Most zooplankton are microscopic but some are larger animals such as jellyfish and jellyfish-like Portuguese man-of-war. Some zooplankton, such as copepods and krill, remain drifters all their lives. Many other zooplankters spend only their earliest stages as drifters. Larval stages of fish, worms, sea stars, barnacles, crabs, and lobsters look nothing like their parents, at first. They grow by eating plankton and then change into forms that either settle to the bottom or develop the physical ability to swim.

Phytoplankton, zooplankton, and their predators form the living component of a complex, open-water ecosystem that includes thousands of species. The dynamic, floating ecosystem of the world of plankton extends across the entire ocean's surface and down into the depths for many fathoms.

Behavior

To stay afloat, plankton evolved many ways to control their position in the water. Spikes and other projections help to distribute a plankter's weight over a large surface area, slowing its sinking. Crab larvae and sea star larvae both have projections. Many organisms such as tiny copepods (distantly related to crabs) and diatoms produce oil, which is lighter than water, to help them float. Air-filled floats help many zooplankton, such as Portuguese man-of-war, stay afloat.

Drifting phytoplankton, floating wherever the currents take them, collect in large numbers wherever water masses of different temperatures and salinity meet. Zooplankton may form aggregates covering many miles. Sometimes, plankton exist in small packets easily missed by plankton nets or by predators. Some zooplankton and small fish migrate towards the surface to feed on phyto-plankton at night.

Ecology

Besides light, other necessary components of this drifting ecosystem include appropriate temperature and adequate *nutrients*. Although the crystal clear waters of some tropical seas such as the Caribbean appeal to human aesthetic sensibilities, the water is clear because it lacks sufficient nutrients to support abundant growth of phytoplankton. If phytoplankton numbers are low, so are zooplankton numbers. The cloudy, murky, green waters of the temperate oceans, particularly in the coastal areas, contain the phosphorus and nitrogen necessary for high productivity in the phytoplankton community. Freshwater runoff from the land provides fertilizer for these rich pastures. A large, diverse community of zooplankton graze on the phytoplankton. Shrimp and other small animals gorge on the zooplankton and so the food web expands to fish, birds, whales, and humans.

Diving Deeper: Leading a Double Life

An overwhelming number of sea-dwelling creatures spend at least part of their lives as zooplankton. Oysters, corals, barnacles, and mussels that as adults are attached firmly to the bottom, colonize new areas by having planktonic larval forms. The attached adults shed their eggs and sperm into the ocean where fertilization takes place, and larvae flow with the currents. Even mobile animals like lobsters, crabs, and worms begin their lives as zooplankton feasting on phytoplankton before settling down on the ocean floor. Many oceanic fish spend their first few weeks or months as lar-vae or juveniles drifting as plankton.

Plankton are Important!

• *Phytoplankton produce 95 percent of the oxygen generated in the sea and one-third to one-half of the earth's oxygen.* All of the oxygen that living organisms breathe is generated by plant photosynthesis, the splitting of water molecules using solar energy. Photosynthesis also removes excess carbon dioxide from the atmosphere.

• *Phytoplankton are the breadbasket of the sea.* Phytoplankton are food for zooplankton, and zooplankton are food for fishes, invertebrates, birds, mammals, and humans. The loss of phytoplankton would result in the loss of most marine life. Our commercial fisheries depend on these healthy, freewheeling drifters. Even herring, eels, and lobsters spend time as drifters.

• *Plankton become petroleum.* As plankton settled to the ocean bottom and were buried, compressed, and heated under the right geological conditions for millions of years, they became petroleum. The hydrocarbons found in crude oil were derived from the lipids of phytoplankton, the remnants of algal blooms millennia ago. Oil is found where plankton was most productive during geologic history, along ancient continental shelves, reefs, and inland seas and lakes. The generation of oil from new plankton sediments continues today but not at a rate that would make oil a renewable resource.

• *Zooplankton include the next generation of many commercially important species such as lobsters, clams, crabs, herring, and other fish.* The challenge for the author of this guide and for you is to help our young explorers develop enthusiasm and understanding for the diversity of plankton and the important role these microscopic drifters play in the survival of life on earth as we know it.

Organization of This Guide

This book is an inquiry- and discovery-based, fun teacher's guide designed to accompany the books *Sea Soup: Phytoplankton* (1999) and *Sea Soup: Zooplankton* (2000) by Mary Cerullo. Where appropriate, a cross-disciplinary approach and a variety of learning styles enrich the student's experience. Activities are designed to foster deeper inquiry into the questions posed in the book. Introductory and ongoing activities are included at the end of this introduction. A culminating section includes a wide variety of activities you may use to bring closure to the whole unit.

The activities include background information, Diving Deeper (sidebars for teachers), objectives, National Science Education Standards addressed by that activity, materials, preparation, and procedures. Questions are integrated into the activities where appropriate. Some activities include reproducible Student Discovery Masters. Extension Activities enrich the student's experience and challenge the student to be creative. The guide includes a glossary (with phonetics) and a list of resources including useful books, websites, and organizations.

The sequence of activities in the guide corresponds to the eight questions asked and answered in *Sea Soup: Phytoplankton.* Most of the questions are relevant to the investigation of zooplankton as well as phytoplankton. The icons highlight important notes revelant to an activity.

Use of This Guide

Sea Soup: Phytoplankton and/or *Sea Soup: Zooplankton* and this teacher's guide may be used as a "stand-alone" unit, or the basis from which to explore other topics. The topic *plankton* is easily incorporated into units on oceans, sea life, pond ecology, environment, whales, or food webs. Some activities involve creative problem solving, designing experiments, analysis of poems, analysis of environmental issues, and role-playing. The teacher may assign relevant sections of the book(s) being used and make specific references to appropriate photographs or illustrations appearing in the book(s).

The books and guide are designed to complement each other in a unique approach to discovering the watery world of plankton.

Underlying Fundamental Education Concepts

In response to the growing concern about the lack of scientific literacy, the education and scientific community has published several important reports and sets of recommendations for what all citizens should understand about science and technology. The studies on science education suggest that it is no longer possible or relevant to try to cover the facts of science. Rather, the reports encourage "inquiry science." Learning how to ask appropriate questions, to search for answers, and draw conclusions that may lead to wise decisions are components of the scientific process. Children are inquisitive by nature, and whether or not a child becomes a scientist, he/she still will need to make wise decisions related to science and technology in his/her daily life, the voting booth, and workplace.

The *National Science Education Standards,* published in 1996 by the National Research Council, and *Benchmarks for Science Literacy,* published in 1993 by the American Association for the Advancement of Science, describe the meaning of scientific literacy and the ideas and skills a person should know upon graduation from high school. Less emphasis is placed on knowing scientific facts; more emphasis is placed on understanding scientific concepts and developing inquiry skills. To accomplish this, emphasis is now placed on exploring fewer topics, but doing so in greater depth using a variety of methods. Children emerge with a deep understanding rather than memorizing facts. Of the changes in emphases suggested by the National Research Council in the *National Science Education Standards* the following are addressed in this teacher's guide:

• Understanding scientific concepts and developing abilities of inquiry
• Learning subject matter disciplines in the context of inquiry, technology, science in personal and social perspective, and the nature of science
• Activities that investigate and analyze science questions
• Investigations over extended periods of time
• Process skills in context
• Using multiple process skills—manipulation, cognitive, procedural
• Science as argument and explanation
• Applying the results of experiments to scientific arguments and explanations
• Management of ideas and information
• Public communication of student ideas and work to classmates

The specific National Science Education Content Standards appropriate to each activity are stated at the beginning of each activity block.

Ongoing Activities to Use with This Guide

Activity 1: Sea Soup Sub

Objective:

In preparation for a dive into *Sea Soup,* the students design and create their own Sea Soup Subs in which to explore the books and the activities.

Introduction:

The realm of the sea, especially the microscopic grassland of phytoplankton and their drifting zooplankton predators, is unknown or at least mysterious to most people. To help untangle the mystery of this important domain, Mary Cerullo, the author of *Sea Soup: Phytoplankton,* uses an imaginary submarine in which the children travel into the miniaturized world of single-cell organisms and their tiny and huge predators, from amoeba-like organisms to whales.

Procedure:

Depending on how much class time you choose to spend on this activity, you may ask each student to draw and paint his/her version of a Sea Soup Sub, or each pair of students may create a three-dimensional Sea Soup Sub using a variety of materials which you or the art teacher assembles.

To begin creating the sense of being under the sea, hang the submarines around the walls of the room or from the ceiling. Additional creations will be added as a result of several activities throughout the Sea Soup unit.

Activity 2: Recipe For Sea Soup

Objective:

The class creates a kettle into which they make contributions to the sea soup as they discover what sea soup is and why it is so important.

Just for your information, here's a Sea Soup recipe:

Ingredients:

- 1,000 gallons of water
- 350 cups of salt
- Pinch of nitrogen, phosphorus, and minerals to taste
- Scoops of "marine snow," bits of detritus (dead material) raining down from the surface
- Billions of tiny phytoplankters (photosynthesizing cells)
- Millions of tiny zooplankters (who love to eat phytoplankters and each other!)
- 1 big spoon for mixing air into the soup

Directions:

1. Mix all the ingredients except the zooplankters together and warm to about 10° C (50° F).

2. Let the mixture sit in the sun or under a huge lamp for several days until it looks green.

3. Add the zooplankters and stir again.

4. Periodically sample the soup to look for larval fish.

5. Watch for large zooplankters such as jellyfish, comb jellies, and Portuguese man-of-war who may eat your sea soup. A large group of comb jellies can eat all the plankton in its path!

6. As the sea soup thickens, herring, anchovies, and other fish appear at the kitchen door to eat the soup.

7. A huge plankton-eater appears—a baleen whale! It will move onto better feeding grounds before all the plankton is eaten.

Procedure:

Use paper to create a large soup kettle and ladle on a bulletin board or draw one on a section of blackboard that can be dedicated to sea soup for the duration of the unit.

Before the students begin reading a *Sea Soup* book ask them:

- What is sea soup?
- Who eats sea soup?
- Why is sea soup important?

List the answers on the board and discuss them. Then, write a summary of the answers on cards and post inside the soup kettle. These answers will probably change as the students' work on the unit progresses. Periodically throughout the unit you will want to examine the answers to the three questions and change the information in the kettle. Challenge the students to remind you when they have learned something to add to the kettle or something that requires that you eliminate something from the kettle.

This activity will also help you assess the students' knowledge about plankton *before* diving into *Sea Soup: Phytoplankton* or *Sea Soup: Zooplankton.*

Class Assignment:

Read the introduction to *Sea Soup: Phytoplankton* or *Sea Soup: Zooplankton* in class or for homework.

An Ongoing Activity That May Start with the Next Activity

Activity 3: Plankton Press

Objective:

Students create a newspaper featuring many aspects of the world of plankton.

Procedure:

Plankton Press is a student newspaper to which all students will make a variety of contributions throughout the Sea Soup unit. The newspaper may take the form of an actual newspaper with different issues published weekly, or a website, or a bulletin board-size newspaper changed or added to frequently.

Throughout the unit you will be reminded to ask students to contribute to the *Plankton Press.* Contributions may range from cartoons and editorials to interviews of Sea Soup critters and discussion of environmental issues.

Have fun as you dive into the watery world of plankton!

Question 1
Are They Plants
or Are They Animals?

OBJECTIVES

In this wide variety of activities, the students will:
- Distinguish plants from animals (and learn that it's not always easy)
- Learn how scientists classify and name living things
- Try to distinguish phytoplankters from zooplankters
- Investigate the effects of fertilizer on phytoplankton growth

The National Science Education Content Standards addressed in these activities include:

- *Standard A:* Science as Inquiry
- *Standard B:* Physical Science—classification, sorting
- *Standard C:* Life Science—functions
- *Standard G:* History and the Nature of Science—address the previously known

How Organisms are Classified and Named

It is human nature to classify and name things. Objects or organisms that share similarities are grouped together. Not all of us choose the same grouping system. In the 1700s a Swedish naturalist named Linnaeus recognized the need to agree on a system of classifying living things. The system Linnaeus started is still used throughout the world. Living things are divided into large groups called kingdoms. Each kingdom is divided into many phyla (singular: phylum), each phyla into classes, each class into orders, each order into families, each family into genera (singular: genus), and each genus into species. The categories are based on similarities and differences. Many characteristics are visible, but others require close scrutiny or even chemical analysis.

The final categories of classification are genus and species. Every organism in the world that has been studied has a scientific name, its genus and species. No matter what language is spoken in whatever country, the scientific name of the organism is the same. Most scientific names have some meaning. Often the species name is the name of the person who discovered the new species. Sometimes the names have a Latin root which indicate something about the organism, such as its color. By convention, the first letter of the genus is capitalized, but the first letter of the species is not. Both the genus and species are underlined or italicized. The common name is the name the local people or general population call the organism.

Diving Deeper: Classifying Isn't Easy!

Categorizing organisms into distinctive groups is not always as easy as it might seem. Until forty years ago, biologists thought there were two kingdoms, plants and animals. Fungi such as mushrooms were included with the plants. However, since fungi do not use the sun's light energy as their immediate source of energy, that is, fungi do not photosynthesize, eventually fungi were put into their own kingdom. Fungi get their food energy by using enzymes to break down other living or dead material. The cell structure of fungi, plants, and animals is sufficiently different to put them in separate kingdoms.

When microscopes allowed the discovery of tiny bacterial cells, scientists were not sure if they were plants or animals. As microscopy and scientific tests improved, bacteria emerged as totally different from plant or animal cells. They lack many of the subcellular structures, such as a nucleus, that exist in plants, animals, and fungi. So now there is another kingdom, Monera, that includes bacteria, blue-green algae, and simple cells, some of which use pigments and chemical systems to photosynthesize or chemosynthesize to make their own food, and others that use food molecules produced by other living things.

About forty years ago, questions arose about tiny photosynthetic organisms such as single-cell Euglena, a "green alga." When conditions are adequate, it photosynthesizes like a plant. When conditions change, it uses its flagellum to swim to find food particles which it takes into its gullet and slowly digests like an animal. So, into what major group or kingdom should Euglena and similar cells be categorized? A new kingdom, the Protists was named. By the 1970s scientists decided that maybe all life could be categorized into five kingdoms: Monera, Protists, Fungi, Plants, and Animals.

In the 1990s scientists investigating the biochemistry of a group of organisms discovered a new organism that is very different from any known cells. As of 1999 biologists generally agree that based on cell structure, function, and biochemistry there are three "domains": Archaea, Prokaryote (the Monerans), and Eukaryote (everything else) into which all living things can be classified. (Note: Viruses are not cells, and are not considered to be "alive." They consist of a protein coat around a clump of genetic material. In order to replicate or reproduce, virus particles must inject their genetic information into living cells. The viral genetic material uses the internal components of the living cell to reproduce the viral genetic material.)

As this guide is being written, a major breakthrough is occurring in the biochemical genetics of plants. The new discovery throws the current classification system into chaos. Scientists have agreed not to start renaming plants yet.

Activity 1: **Class Classifies Critters**

Objective:

Using an odd assortment of objects, students devise systems of classifying "organisms."

In this investigation students explore the similarities and differences between things, and learn how scientists identify, categorize, and classify organisms from each other.

Materials:

For each pair of students:
- 1 box of nuts and bolts
- 2 copies of Student Discovery Master 1: Nuts and Bolts

For the class:
- 1 bag of a variety of snack food packages

Note: Each box of nuts and bolts should include about a dozen different, small objects such as screws, nails, nuts, bolts, and tacks of assorted sizes. Snack foods could include potato chips (no-fat or baked; fried; baked or fried and flavored; not flavored; ridges; no ridges; etc.), and corn chips (baked or fried; blue or yellow; salted or unsalted; etc.)

Procedure:

1. The activity is designed to help students understand how things can be categorized and classified.
- Begin the class by pulling a variety of packaged snack foods out of a large bag. Encourage the class to discuss the similarities and differences among the objects. How many different ways can they group the snack foods? Should the most obvious difference always be the first difference used to separate the groups? Do we commonly categorize objects in our lives? Why?
- Quickly point out how scientists categorize organisms. For example, the easiest way is to divide the group of *organisms* into two different groups based on one distinguishing characteristic (corn vs. potato chips). Then take one of the two groups and again divide it into two different groups based on another distinguishing characteristic (baked corn chips vs. fried corn chips). Each of these groups is then divided (fried, blue corn chips vs. fried, yellow corn chips). Again subdivide (fried, yellow, flavored corn chips vs. fried, yellow, non-flavored corn chips). One final category might be fried, yellow, chili-flavored corn chips; fried, yellow, nacho-flavored corn chips; fried, yellow, lime-flavored corn chips.

2. Introduce the activity by explaining that the students are beginning a fun journey exploring how taxonomists (scientists who study the classification of

things) work.

• Divide the class into working pairs. Distribute a copy of Student Discovery Master 1: Nuts and Bolts to each student, and one container of nuts and bolts to each pair. Allow about twenty minutes for the students to sort and categorize the objects as many ways as they can. They could categorize according to size, shape, function, color (kind of metal), sharpness, etc. Have each pair share with the class one way that they categorized the objects. (They may lack the vocabulary to name the object.)

• Close the activity by leading a discussion about how animals are categorized and classified. What obvious trait is used to separate vertebrates from all other groups? What traits are used to separate fish, amphibians, reptiles, birds, and mammals from each other? Why do we need to classify organisms? How does it help scientists understand animals better? How does it help identify an organism when you don't know what it is? This might be an appropriate time to introduce the concept of scientific nomenclature, i.e. scientific names for all different types of organisms.

Extension Activity:

Students may be interested in how things are named. Encourage them to look up "Scientific Names" in science books or the Internet. A scientific name is the name given by a taxonomist to one kind of organism. Refer back to the background information above.

Activity 2: What Makes a Plant, a Plant, and an Animal, an Animal?

Objective:

Students will share their experience and knowledge in setting criteria for categorizing a living thing as a plant, animal, or other.

Procedure:

1. Allow the children time to read the question, "Are they plants or are they animals?" in the book, *Sea Soup: Phytoplankton* or an appropriate section of *Sea Soup: Zooplankton.*

2. Organize the class into teams of three children.

3. Give each team three copies of the handout, Student Discovery Master 2: Plant or Animal?

4. Explain the purpose of this activity (elaborate on the objective stated above).

5. Allow time for the students to do the activity.

6. With the children remaining in their groups, lead a class discussion of the lists. Write a composite list on the board and then create a Venn Diagram using the class data.

If they have omitted any significant traits, lead them towards thinking of those traits by asking leading questions such as:

• If animals eat food for energy to grow, reproduce, and move, how do plants get energy to grow and reproduce?

• Do plants move?

• If you put plants and a small animal in a closed container with some water and food, the plants and animal usually live. What does the plant give off into the air that helps the animal breathe? Do animals give off anything into the air that helps plants live?

• Do plants breathe? Do they have lungs or gills?

Try to emphasize that classifying and categorizing organisms is not always as easy as we might think initially. The students may begin to realize that plants and animals (and living things in general) share a lot of common characteristics.

Activity 3: Is This a Zooplankter or a Phytoplankter?

Objective:

Students experience the challenge of sorting plankton at the most basic level, that is, separating plants (phytoplankters) from animals (zooplankters).

The goal of this activity is to realize that zooplankton and phytoplankton are very similar in appearance. The students may have great difficulty accurately sorting the images. A point to emphasize upon completion of the sorting is that sometimes traits other than appearance may separate groups of organisms. For example, phytoplankton photosynthesize, zooplankton do not. Unless you can see photosynthetic pigment in the plankters, you may not be able to distinguish between phyto- and zooplankton.

Sorting plankton under the microscope is an interesting and challenging task. Many laboratory scientists spend hours doing just that! What's in a sample and how many of each organism are there?

Note: You may have noticed some variations in the word "plankton." "Plankton" refers to many planktonic organisms. "Plankter" refers to one, while "plankters" also refers to many or a collection of plankton.

Materials:

• Copies of the Student Discovery Master 3: Sort a Plankton Sample (one set per group). Note: Be sure to block off the answer key when making copies for the class!)

• Transparency copy of the above cut into cards

• Scissors—one pair per student group

Procedure:

1. Divide the class into groups of two to three and give each group a set of Student Discovery Master 3: Sort a Plankton Sample.

2. Students cut the pages into separate cards.

3. Challenge the class to sort their cards into two categories, animals and plants.

4. After five to ten minutes, encourage the students to move around the room looking at the way others sorted their plankton sample.

5. Now, use the overheads to show the actual groupings, and encourage the class to argue why they might disagree with the groupings. For example, ask if any pair will state why they did not include specimen A as a plant. What characteristics of an animal or plant does that specimen illustrate? (Depending on your class, you might like to keep a tally of how many pairs categorized each of the twelve organisms as a plant, or as an animal. This will clearly illustrate why classifying organisms is not always an easy task even for professional scientists who specialize in taxonomy.)

For class discussion or a written assignment (homework):

1. Which specimens were the most difficult to categorize?

2. Is size a good indicator of whether something is a plant or an animal?

3. Are visible characteristics always good indicators of how organisms are related? Explain your answer.

4. What other characteristics of organisms might help in determining whether they are related?

5. Give an example of two organisms that look like they are closely related, but actually are not. (An example is a black bear and a panda, or a whale shark and a baleen whale.)

Activity 4: What Will I Grow Up to Be?

Objective:

Students try to match plankters with the adult form of the critter.

Background:

Plankton contains multitudes of different kinds of organisms: single photo-synthetic cells; larval forms of sea stars, crabs, lobsters, worms, fish, snails, clams, oysters, and young shrimp. Many marine biologists study plankton. They gather samples and examine the samples under the microscope to determine what species are present and how many of each species are in the sample. Using the plankton data samples from many areas and data taken from the same areas but during different seasons or year to year, they can then compare the data and note changes in diversity and population size. It helps fisheries biologists to predict whether the oyster set (larval oysters settling and attaching to rocks and shells) will be good that year. Or, the lobster catch in eight years can be predicted by the relative number of lobster larvae in the plankton samples. The number and type of plankton also are good indicators of the food availability for the organisms which prey on plankton such as herring, basking sharks, and jellyfish. The impact of major freshwater floods on the population of organisms in an estuary can be measured by sampling the plankton. However, to identify the groups to which the organisms in the plankton sample belong requires training, perseverance, and skill.

Your class can begin to appreciate the training necessary by trying to match the planktonic organisms in our "sample" with the illustrations of adult marine organisms.

Materials:

• Copy for each student Student Discovery Master 4: What Will I Grow Up to Be?

Note: Be sure to block off the answer key when making copies for the class!

Procedure:

1. Review the question, "Plant or Animal?" and discuss what other types of plankton exist besides phytoplankton. Lead the students in discussing that many animals in the sea reproduce by shedding eggs and sperm into the water where fertilization occurs. The fertilized eggs grow into tiny larvae and become part of the zooplankton. The animal plankton (zooplankton) feed on phytoplankton. When the animal larvae in the zooplankton grow large enough, they change form again and become adults. Sometimes this means that the larval stage changes into a sedentary stage. An example is an oyster that swims in its larval stage, and is stuck on a rock or shell as an adult. Only a few eggs survive to adulthood. Most eggs and larvae are eaten by plankton feeders. Larvae are a very important part of the food web.

2. Hand each student a copy of Student Discovery Master 4: What Will I Grow Up to Be? Let the students discuss among themselves which critter grows into which adult. Chances are that they will have to guess many of them because the plankter looks so different from the adult.

3. Point out that larval forms can look very similar to each other, while their adult forms are very different. Why might larval forms be so similar? Could a hard-shelled tiny clam float as plankton as well as the soft larval form? Could a sea star larva float if it was covered with a tough, spiny coat like its parent? Could it swim? Does a lobster larva need claws?

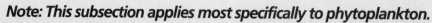

Photosynthesis

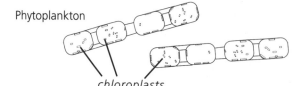

Phytoplankton

chloroplasts

Background:

The most critical living component of sea soup is phytoplankton. Within the tiny cells, the energy of sunlight is converted to chemical energy as the cell makes sugars and starches, the source of energy for the animals of the open ocean. Because of their photosynthetic activity, phytoplankton are the most important producers of organic matter in the ocean. Animal life depends on them as the primary food source.

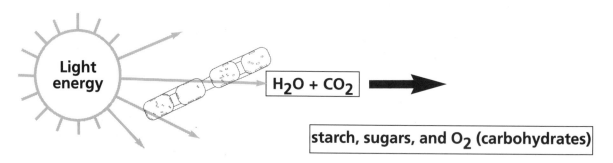

Light energy

$$H_2O + CO_2 \longrightarrow$$

starch, sugars, and O_2 (carbohydrates)

The process of converting light energy to chemical energy is called *photosynthesis*. It occurs with the help of pigments in the cells that absorb light. Water molecules and carbon dioxide break apart and reform to produce carbohydrates (usually sugars and starches) and release oxygen. In order to photosynthesize, phytoplankters require enough light, appropriate temperature, and certain nutrients.

22

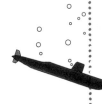

Diving Deeper: **More About Photosynthesis**

During photosynthesis pigments such as chlorophyll and carotenoids absorb light energy, trap it, and use it to break carbon dioxide molecules and water molecules apart. A complex series of chemical steps results in the manufacture of energy-rich compounds such as starch and sugar. Starches and sugars are used by all living organisms in another complex chain of chemical reactions to produce very high energy molecules called ATP which are then used along with other nutrients to build complex proteins, fats, and carbohydrates. Ultimately, the organisms grow, maintain themselves, reproduce, and move because the tiny phytoplankton cells converted the energy of sunlight to usable chemical energy.

Phytoplankters photosynthesize only within the proper range of light intensity, temperature, and salinity conditions. In addition, just as a garden needs fertilizer, phytoplankters need certain nutrients.

• The required nutrients are nitrogen (usually combined with oxygen as nitrate), phosphorus (present as phosphate), carbon, potassium, sodium, calcium, and sulfur, all plentiful in sea soup. Trace metals such as iron, manganese, cobalt, copper, nickel, tin, and zinc, and vitamins, especially vitamin B12, contribute to the photosynthetic process, too. Some of these nutrients may be available in such small quantities that they are called "limiting nutrients." Photosynthesis will slow down or cease when the available supply of a nutrient decreases to a certain level. Nitrogen is usually the limiting nutrient in marine phytoplankton. As you will discover later in this guide, sudden availability of a limiting nutrient may account for sudden algal blooms such as red tides.

• Terrestrial photosynthesizers receive a full spectrum of light wavelengths ranging from ultraviolet through the visible spectrum to infrared wavelengths longer than we can see. However, life for photosynthesizers beneath the ocean's surface is a different matter. Water absorbs and filters light removing much of its energy and changing its color. Suspended and dissolved materials in the water further affect the color and intensity of light passing through the water. Certain wavelengths are absorbed or scattered more than others. Clear sea water in the open ocean absorbs most red and violet wavelengths within 75 feet of the surface. Blue light transmits better than any other color. (This explains why SCUBA divers see the underwater scenery bathed in an eerie, blue light.) In coastal areas, dissolved and suspended matter removes much of the blue light. Light of all wavelengths is absorbed quickly. Two-thirds of the available surface light is absorbed in the first 3 feet of water and 95 percent within the upper 35 feet. Yellow-green light predominates just below the surface.

• How does this relate to photosynthesis? Chlorophyll, the primary pigment in photosynthesis, absorbs and harnesses red and violet wavelengths most efficiently. Several kinds of chlorophylls have evolved to absorb slightly different parts of the spectrum. But, as stated earlier, very little violet and red light penetrates much below the ocean's surface. To compensate for the difficulty in absorbing sufficient light, multicellular marine algae and phytoplankton evolved accessory pigments that assist their chlorophylls in collecting sunlight in the challenging underwater light environment. Chlorophyll absorbs reds and violets and reflects unusable green wavelengths. Therefore, cells containing just chlorophylls appear green to our eyes. Accessory pigments such as carotenoids absorb blue and reflect yellow. Other accessory pigments absorb blue and green light. Depending on the combination of pigments, algae and phytoplankton may range in color from green to red to golden-brown.

Extension Activity: Develop a Mural

You may choose to have your students develop a mural answering the question, "Why can't plants live in deep water?" (Show how light fades as it moves into deeper water.)

Activity 5: The Effects of Fertilizer on Phytoplankton

Objective:

The class or student teams design an experiment to test the effect of fertilizer on the growth of phytoplankton.

Background:

Just as terrestrial plants grow more rapidly under ideal conditions including appropriate temperature, water, light, and *nutrients* (fertilizer), phytoplankton also respond by rapidly multiplying when conditions are ideal. However, sometimes too much of a good thing—that is, too much fertilizer—may result in the death of the plant or phytoplankton culture. In this investigation, students design the experiment within certain parameters established by you. *(Note: This experiment will run for the duration of the Sea Soup unit.)*

Procedure:

1. Read through this whole section and then decide if the experiment will be designed by the whole class, or if teams of students will design their own variation of the experiment.

2. Depending on the location of your school, you may choose to gather pond water or marine phytoplankton yourself, or, to ensure the presence of healthy phytoplankton, you may order cultures of marine or freshwater phytoplankton from a biological supply house.

3. Start with nutrient-rich water in a gallon jar (freshwater or saltwater depending on your plankton culture).

4. Inoculate the water with several droppers of a mixed phytoplankton culture.

5. Mark the water level on the outside of the jar.

6. Place the jar in a well-lighted area but not in direct sunlight (a place where measurements of cloudiness of the water can be measured without moving the jar).

7. Observe the clarity of the water. Is there any color to the water? If so, what color?

8. Make an "X" about 1/4" tall in the middle of a 5" x 8" card.

9. Place the card against the far side of the jar. The students should be able to read the "X" on the card. Now, move the card farther away from the jar until the students can no long read the "X." Measure this distance.

10. Twice a week students should gently add distilled water to bring the water level up to the mark.

11. Twice a week students should measure the turbidity with the card.

12. Record the data on a class data sheet and make a graph plotting the distance of the readable card from the jar versus the number of days since the experiment started.

13. Is there a change in the density of the phytoplankton?

14. Did the density of phytoplankton reach a peak and then decrease? If so, explain what factors might have caused that to occur.

Extension Activity: Design an Experiment to Test the Effect of Fertilizer on Phytoplankton

Objective:

Students will work in small groups to design and conduct a long-term experiment on the effects of fertilizer on phytoplankton growth.

1. Students may be aware of the use of fertilizers by farmers and homeowners. Lead them in a discussion about fertilizers.

• What are fertilizers? (They are nutrients necessary for a plant's healthy growth. In natural ecosystems, the soil contains nutrients necessary for plant growth. The plants that can grow there are adapted for life using the nutrients in the soil.)

• Where do fertilizers come from in the sea? (Naturally occurring minerals dissolve into ocean water. Water runoff from land carries nutrients into rivers, estuaries, and the sea.)

• Commercial fertilizers contain the basic nutrients nitrogen and phosphorous. Why are they used? Will the plants grow without fertilizers? Why or why not? (Fertilizers usually help plants grow bigger and faster than they grow using just the nutrients in the soil.)

• Do natural ecosystems such as a forest or a wetland need fertilizers? If so, where do they get fertilizer? (The nutrients recycle back into the soil as the dead plant matter decomposes. Growing plants then use the recycled nutrients to build new cells and grow.)

• What happens to a forest's ability to regrow as well as it once did, if it is clear-cut and all the plant matter is carried away?

• What does crop rotation mean? (Alternating kinds of crops grown in one area. Some plants can take some nutrients such as nitrogen out of the air and "fix" it so that it is usable by other plants.)

2. Discuss hypothesis formation and controls with your class.

• What hypothesis could be tested using the method of measuring growth described above?

• What procedure would help test the hypothesis? (Set up two additional jars that have been seeded with the same amount of phytoplankton. To one jar, add one quantity of liquid plant fertilizer. To the other jar, add twice that amount of liquid plant fertilizer. Make observations on the turbidity of the three cultures twice each week.)

• What is the control?

If you choose to have the students work in groups use *Student Discovery Master 5: Effects of Fertilizer on Phytoplankton Growth.*

Diving Deeper: **The Scientific Method**

All scientists from students to Nobel Prize winners use a similar approach in their investigations: the scientific method. Something spurs their interest. After some observations are made (and maybe something of interest is read), a testable question comes to mind. The question is rephrased as a hypothesis (an educated guess). The hypothesis predicts the outcome of an experiment or observational study. A procedure is planned for gathering data. This is called the experimental design. After gathering and analyzing the data, the scientist draws conclusions. Do the data support or reject the hypothesis?

1.   Make preliminary observation
2.   Create a testable question
3.   Formulate a hypothesis
4.   Organize a procedure
5.   Gather data
6.   Analyze data
7.   Draw conclusions

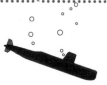

Student Discovery Master 1:
Nuts and Bolts

Name _____ **Date** _____

Class _____ **Names of partners** _____

Challenge:

To sort the items in the box by similarities and differences. This activity shows how taxonomists (scientists who study how organisms are related to each other and who give scientific names to all living things) begin their work.

Materials:

• Box of nuts and bolts

Procedure:

1. Examine all the "specimens" as if they were unknown creatures just brought back from the tropics.

2. Arrange them into groups that share similarities. How many categories do you have? _____

3. Now, lump the objects together so you only have *two* categories. What characteristic separates the two groups?

4. Now, begin to subdivide the objects in each of the large groups into smaller and smaller categories. Create a diagram to show how you categorized the objects.

DIAGRAM

5. Have you subdivided so that finally each *organism* is in its own category? If so, then each *organism* is a separate species.

6. When you have finished, turn over the paper and see if you can create another classifying system. Draw it.

Student Discovery Master 2: Plant or Animal?

Name _____ **Date** _____

Class _____ **Names of partners** _____

Challenge:

To think about how animals and plants are the same and how they are different.

Procedure:

1. Discuss the characteristics of plants and animals with your group. List them in two columns on this paper.

2. Now, put a check next to characteristics that are the same for both plants and animals.

3. If most of the traits are checked, then review your lists and try to add more specific, descriptive, distinguishing traits for animals, and then for plants.

4. Fill in a Venn Diagram below to show the similarities and differences between plants and animals. (When you are done, your teacher will lead a class discussion of this topic.)

5. Compare your lists and Venn Diagram with the one on the board.

6. In a different color pencil or pen, add characteristics to your diagram that you had not included before the class discussion.

VENN DIAGRAM

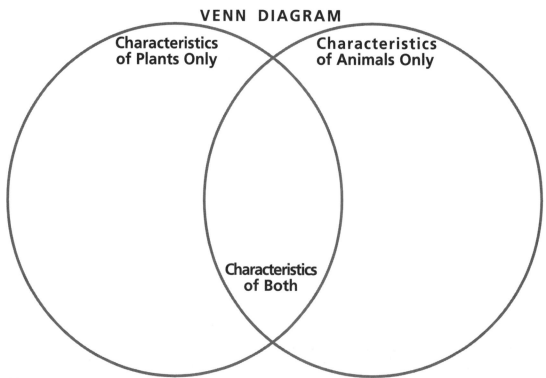

28

Student Discovery Master 3:
Sort a Plankton Sample

Name _____ Date _____

Class _____ Names of partners _____

A.

B.

C.

D.

E.

F.

G.

H.

I.

J.

K.

L.

ANSWER KEY: A. zooplankton B. phytoplankton C. phytoplankton D. zooplankton E. phytoplankton F. phytoplankton G. phytoplankton H. zooplankton, I. phytoplankton J. phytoplankton K. phytoplankton L. phytoplankton

Student Discovery Master 4:
What Will I Grow Up to Be?

Name _____ **Date** _____

Class _____ **Names of partners** _____

Challenge:

Match the plankter to the adult it will become!

Plankton includes many young or larval forms of animals that grow and change into adults that look very different from the larvae. Most planktonic larvae look somewhat alike because they have similar life styles. They are tiny and they can only eat tiny things. They have to stay up in the water. If they sank, they would not find food.

Discuss with other students what plankter grows into what adult. When you have made a decision, draw a line between the two. Continue until you have matched all the zooplankters with the adult forms.

Your teacher will go over this with the class.

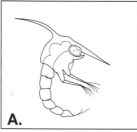

A.

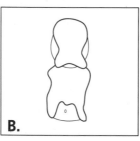

B.

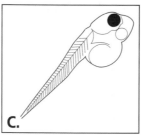

C.

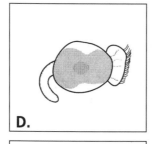

D.

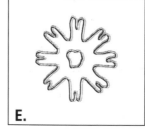

E.

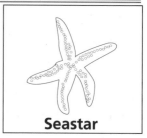

Seastar

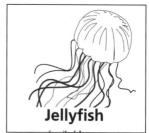

Jellyfish

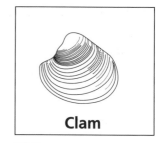

Clam

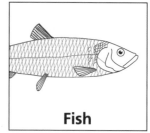

Fish

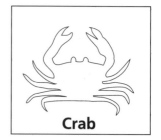

Crab

ANSWER KEY: A. Crab   B. Sea star   C. Fish   D. Clam   E. Jellyfish

30

Student Discovery Master 5:
Effects of Fertilizer on Phytoplankton

Name _____ **Date** _____

Class _____

Challenge:

Design and conduct a long-term experiment to test the effects of fertilizer on plant growth.

Phytoplankton grow and reproduce by using the sun's light energy to make sugars, starches, and other substances the plants need to survive. They store those substances and use them when they grow and reproduce.

What are the requirements for a plant to grow? Check your answers with your teacher.

Before an experiment is done, the experimenter must ask a question to be tested by the experiment. Then, the question is worded as a sentence that predicts the outcome of the experiment. Usually the experimenter knows something about the materials being used and tested in the experiment, so the prediction is based on an informed guess about what the experiment might show. The prediction is called a *hypothesis*.

All experiments must have a *control*, something with which to compare the results of your experimental manipulation.

State the question you are asking in your experiment.

Now restate the question as a hypothesis.

List the materials needed to conduct this experiment.

What is your control?

List the steps in the procedure you will follow in the experiment.

Design a table for recording your observations (data).

Hand this form into your teacher. Your teacher will evaluate it and return it to your group and discuss it with you.

Now, set up your experiment. Make a half-page table on which you will record your observations. After your observation time has passed, summarize your data, graph your data, and write your conclusions.

Summary:

Graph: (Attach a separate page with your graph.)

Discussion and Conclusions:

Application:

Predict the impact of nutrient pollution on phytoplankton. What outcome would you predict if fertilizer runoff from a farmer's cornfield washes downstream into the ocean?

Question 2
Can Plankton Swim?

By definition, plankton are organisms living somewhat suspended in water and carried by currents and waves. Plankton are incapable of significant locomotion against the action of the water. But not all plankton are microscopic or small. For example, some large jellyfish are considered zooplankton because their swimming movements barely keep them oriented upright. Tiny zooplankton and some phytoplankton are capable of movement and may actually migrate up and down in the water column, but they cannot move significant distances horizontally. So, the question is, How are plankton designed to survive? How do they manage to "float" and not sink to the bottom? What shapes work best to help them survive? What diverse forms have plankton evolved in order to survive?

OBJECTIVES

In these activities students explore specific examples of phytoplankters and how plankton, in general, are designed.

The students will:

- Create edible models of three types of phytoplankters or zooplankters.
- Create a classroom demonstration of relative sizes of phytoplankton to zooplankton and other ocean dwelling organisms.
- Discuss and investigate the relationship between surface area and volume.
- Design a phytoplankter or zooplankter.

The National Science Education Content Standards addressed in these activities include:

- *Standard A:* Science as Inquiry
- *Standard B:* Physical Science—properties of objects (density)
- *Standard C:* Life Science—characteristics of organisms; structure and function; diversity and adaptations of organisms

Note: If you are only using Sea Soup: Zooplankton, modify Activity 1 by having students create models of zooplankton based on photographs and information in the book.

Activity 1: Who Are the Phytoplankters?

Objective:

Students create edible models of three types of phytoplankters.

Background: *Sea Soup: Phytoplankton* emphasizes three types of phytoplankton organisms. All three are members of the kingdom Protista.

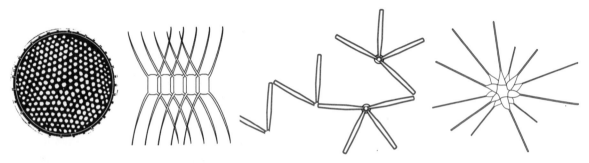

Diatoms are probably the most important food source in the ocean. They are eaten by small zooplankters and by larger critters such as oysters and clams. Their role as primary producers (photosynthesizers) at the bottom of the food chain also makes their contribution to the earth's oxygen immensely important. These flat, plate-like cells are very common members of phytoplankton. Each cell has a two-part wall, like a lid on a tie box.

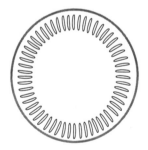

Coccolithophores are tiny spheres covered with numerous chalky discs called coccoliths. When the cells shed the coccoliths, the ocean turns white. The coccoliths sink to the ocean floor, are buried, and are either transformed into limestone or broken down and then used by coral animals to create the hard coral structure. The White Cliffs of Dover (England) are made of coccoliths.

Dinoflagellates rank second to diatoms in importance as marine primary producers. Many species of dinoflagellates luminesce and glow in the wake of a boat and in breaking waves. Spines, wings, and horns may decorate the cell wall. A single, long flagellum lies in a groove around the middle of the cell and another flagellum lives in a groove perpendicular to the first one. This flagellum gives the cell some limited mobility as it thrashes about in the water. Dinoflagellates are responsible for many of the "red tides" that may poison fish and humans. (This is discussed in Question 6 Background.)

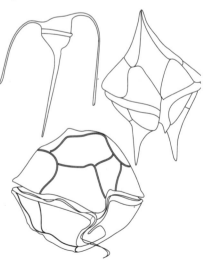

34

Materials:

- Marshmallows
- Marshmallow Fluff
- Bananas
- Apples
- Raisins
- Toothpicks
- M & M candies
- Cake-decorating gel in tubes
- Vanilla wafers
- Licorice whips
- Bread
- Cheerios
- Cookie cutters
- Peanut butter
- Sprinkles
- Paper towels
- Dull butter or dinner knife
- Drawing paper
- Crayons, markers, pencils

Procedure:

1. Assemble all materials.
2. Set up identical stations at which teams of three students will work.
3. Allow time for the class to read the "Can phytoplankton swim?" section in *Sea Soup: Phytoplankton* or the appropriate section of *Sea Soup: Zooplankton,* and examine the photomicrographs, looking at the variety of design in plankton structure.
4. Divide the class into teams of three students.
5. Ask the students to sketch and label three types of phytoplankters on the drawing paper.
6. Now, the students use their creative ideas in building models of the three types of phytoplankton using some of the edible materials supplied.
7. Rules include no eating—yet!
8. When the students complete their models, they should place them on towels.
9. Before they eat their models, ask the students to circulate around the room to look at all the models.
10. Can they guess what organisms the models represent?
11. Have everyone stand who made diatoms. Ask them to tell the class something about diatoms.
12. Repeat for dinoflagellates and coccolithophores.

Extension Activity or Homework: Definitions

You may choose to have the students start a glossary/spelling word notebook that they add to throughout the Sea Soup unit.

Read or copy the quote from *Alice in Wonderland* (in the introduction to this guide) to the class. Ask them to look up words they do not know (tureen, dainties, stoop). How might this poem relate to what we are studying now? Why is the soup green? What or who are "the Dainties"? What or who might stoop to eat "such dainties"? Students who are truly motivated and curious might choose to borrow a copy of *Alice in Wonderland* from a library and look up the quote to see how it fits into the story.

Diving Deeper: A Few More Details

*D*iatoms contain green chlorophyll and yellow carotenoid pigments. As a consequence, diatoms often appear yellow. They are usually single celled but frequently occur in stacks or chains of cells. The cell shape of different species varies. Generally, planktonic diatom cells are flat and plate-like (radially symmetrical). Elongated diatoms (bilaterally symmetrical) move by gliding. They are most often found on a solid substrate such as rocks or large algae, and are not usually free-floating. All diatoms have an external cell wall or test composed of silica, a glass-like substance. The test has two parts; a larger half fitting snugly over the slightly smaller half, like a lid on a box. Exchange of materials, such as nutrients acquired from the surrounding water, occurs through tiny pores or holes arranged in fine lines on the test's surface.

Coccolithophores are in the same taxonomic group as diatoms but look very different. Coccolithophores, "plants in armor," usually have 12 plates or coccoliths so small that 1,500 plates could fit on the period at the end of this sentence. Coccoliths are made of calcium carbonate like our teeth and bones.

Dinoflagellates contain pigments similar to those of diatoms. They usually have a cellulose cell wall perforated by many pores. Dinoflagellates resemble both plants and animals—a good example of the quandary scientists have when classifying organisms. The cells photosynthesize, but they also swim and they eat other plankters when they cannot photosynthesize. Some dinoflagellates produce strong neurotoxins, poisons that affect the nervous system. When a bloom of dinoflagellates occurs, their bright-red stored oils may discolor the water, forming a red tide.

Activity 2: The Scale of Things

Objective:

To create a presentation illustrating the relative size of phytoplankters and zooplankters compared with many types of organisms in the ocean, all of which depend on phytoplankton. The challenge is how to use the same scale when representing a whale and a diatom, if possible.

Background:

The size range of organisms considered as plankton is enormous—from ultramicroscopic coccolithophores to 30-meter- (98-feet-) long jellyfish. Diatoms and dinoflagellates cells range in size from about 0.015 millimeter (0.0032 inch) to 0.1 millimeter (0.004 inch). The weight of phytoplankters spans a range of one millionfold, the same range that encompasses all terrestrial mammals from mouse to elephant. The following fact might help students grasp how small single phytoplankters are: It would take about one trillion large phytoplankton cells to occupy the same volume as an average six-foot-tall human.

Sometimes it is difficult to grasp the importance of microscopic phytoplanktonic organisms to life on earth. Most life on earth ultimately depends on the ability of phytoplankton to photosynthesize and release oxygen to the water and atmosphere, recycle nutrients, and supply food energy for the vast oceanic food web.

In the sea, there are many advantages to being small, and few to being large. Although the blue whale, the largest animal ever to have lived, lives in the sea, on the whole marine organisms are much smaller than those on land. A phytoplankter has no need for roots, stems, trunks, branches, leaves, or flowers. The adaptations of phytoplankton are simple: small size to absorb nutrients from their surroundings and slow sinking; spines to slow sinking and to discourage predators; and the ability to reproduce rapidly under favorable conditions, as well as the ability to endure when the conditions are not favorable.

Each zooplankter that eats phytoplankton must be the right small size to obtain its particular minute food from the sea soup in which all plankton are suspended. They need some method for feeding on phytoplankton or zooplankton. Zooplankters range from microscopic snail larvae to 1.5-inch long krill, and larval fish to huge Portuguese man-of-war.

Materials:
- Many balls of string or yarn
- Chalk
- Measuring sticks
- Paper
- Markers

Procedure:
1. Make eight copies of Figure 1.
2. Divide the class into nine groups. Give each group a copy of Figure 1.
3. Assign one organism or group of organisms to each student group.
Note: We cannot represent all these organisms to scale because diatoms, dinoflagellates, and coccolithophores are microscopic, with coccolithophores being so small that they are only visible with special electron microscopes. Students will probably want to represent the whale shark as full size. Have students discuss a plan for outlining the organisms—the hallway, a black-topped area, or the sidewalk around the school may be an appropriate place to use chalk to illustrate the food web scale.
4. Each group outlines the organism to scale with string or chalk on the floor or playground. For the microscopic organisms, the students may choose to write an explanation on a card taped next to tiny dots!

This activity will help the food web activities in the next section.

Figure 1: Size of Organisms of the Sea

Diatoms
0.0015mm
(0.0032 in)

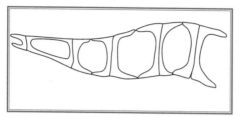

Dinoflagellates
0.1 mm (0.004 in)

Coccolithophore
ultramicroscopic

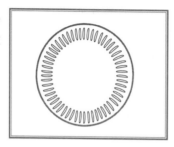

Copepods
1 mm (0.04 in)

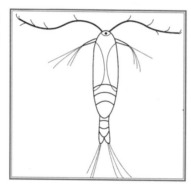

Larval crab
3 mm (.25 in)

Euphasid shrimp or krill
2 cm (1 in)

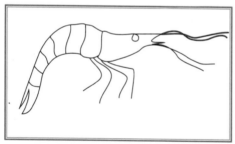

Larval sea star
5 mm (.3 in)

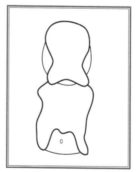

Herring
15 cm (6 in)

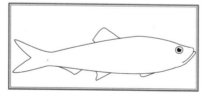

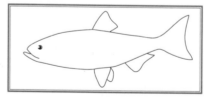

Salmon
1 m (3 ft)

Killer whale
5 m (16 ft)

Whale shark (eats plankton)
17 m (55 ft)

Objectives:

Students will:

• Consider possible mechanisms by which phytoplankton and zooplankton remain suspended in the ocean water. (How do plankton slow down the rate at which they sink?)

• Design a phytoplankter or a zooplankter.

Background:

Marine life is divided into three categories based on lifestyles. Bottom-dwelling organisms, such as crabs, are called the benthic organisms. Strong-swimming animals that live in the open water, such as tuna fish, whales, and squid, are called nekton. Plankton are floating or feebly swimming organisms. All types of plankton are at the mercy of tides, currents, and waves for transportation. Because they require light for photosynthesis, phytoplankton need to be near the surface. Zooplankton that feed on phytoplankton need to be wherever the phytoplankton are.

To stay alive all plankters must stay afloat, but many plankters are heavier than water. Without compensating mechanisms they will sink to a dark death. All plankton species have evolved adaptations for staying off the bottom at least long enough to reproduce themselves. Individual phytoplankters only live for a few hours to a few days. Zooplankters live weeks to years. Many zooplankters change (metamorphose) into adults and settle on the bottom or other hard surface, or swim away into the sea.

The greater the surface area of an organism relative to its volume, the greater its friction as it moves through water, and so the slower it sinks. One way to increase relative surface area is to grow extensions from the cell's surface. Spikes and other projections on the organism help to distribute the organism's weight over a large surface area, slowing its sinking, e.g., sea star larva. Just being small makes the surface-to-volume relationship advantageous.

Some cells have a large cavity at their center. This reduces their mass and density relative to their surface area. In many phytoplankters, stores of oil and fat give them buoyancy.

Materials:

- 1 cup cooking oil tinted with food coloring
- 1 cup water tinted with another food coloring
- Clear glass beaker or jar
- Metal or other top from a jar (at least two inches in diameter and capable of sinking)
- Ball of clay equal in weight to the jar top
- Scale
- Gallon jar or aquarium filled with water
- Pile of pieces of 8 1/2" x 11" paper
- Scissors or ruler

Use two demonstrations to show how the plankters remain in the water column (i.e., why they do not sink).

Procedure:

1. Select two students to help you demonstrate. Show the class the two cups of colored liquids. Ask the two students to carefully pour the liquids into the clear glass beaker or jar. Ask the class why the liquids did not mix; one floated on the other. Discuss density (weight per unit volume). Explain that one of the liquids (oil) is less dense than water and therefore it floats on water. How does this relate to plankton?

2. Next, hold up the ball of clay and the jar top. Ask the class to predict which will sink the fastest in water. (The clay ball.) Then ask them, Why? If they say because it is heavier, show them with a simple balance or tell them that they weigh the same. Now drop both objects into an aquarium filled with water. What happened?

3. Now, challenge the class to design a simple test to determine how body shape affects sinking using just two pieces of paper. (Here's how to do it. Tear a sheet of paper into two equal pieces. Squeeze or roll one piece into a ball. Hold one piece in each hand and extend your arms to shoulder height. Get ready to observe the movement and rate of falling. Drop both pieces at the same time. What happened?)

4. Discuss how the shape of an organism affects its rate of sinking.

Assignment (in class or homework):

Design Your Own Phytoplankton or Zooplankton

1. When designing a planker, consider the following questions:
   - How will it stay afloat in the water column?
   - What shape will it be?
   - Will it be able to swim? How?

2. Be creative! Maybe your organism sinks to the bottom at night and swims up to the light during the day. Maybe it acts like an animal at night and catches prey, then photosynthesizes during the day like a plant.

3. Remember, most phytoplankton end up as food for zooplankton. Any ideas on how a phytoplankter could avoid being eaten?

4. Draw your plankter.

5. Label the parts and explain their functions.

Extension Activity: Name Your Plankter

• Give it a common name and a scientific name. Use a dictionary or root-word book.

• Build a three-dimensional model of the plankter to hang from the classroom ceiling.

Diving Deeper: Diatoms Sink

Phytoplankton cells need to be near the surface in order to receive sunlight for photosynthesis. However, many of the nutrients the cells need are below. A cell cannot float indefinitely at the surface, but must remain in motion or it may deplete the nutrients from the surrounding sea soup. Motile phytoplankters may move into nutrient-rich waters. Non-motile cells, such as diatoms, sink at an average rate of one meter per day. This is slow enough to allow the cell to divide several times before leaving the lighted zone. Diatoms can only thrive when assisted by vigorous stirring by currents or upwelling.

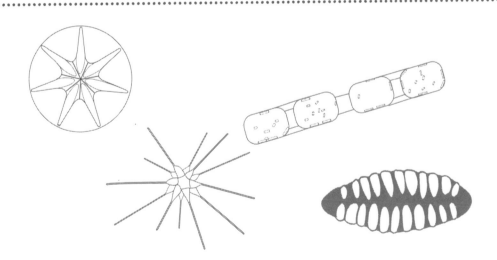

Question 3
How Many Phytoplankton Does It Take to Fill a Humpback Whale?

Scientist John H. Ryther said, "It is the fate of most phytoplankton to be eaten." Phytoplankton are the base of the food chain in the sea. Zooplankton and everything else in the sea feeds on them directly or indirectly. A few escape being eaten only to die and sink into the ocean's mud and eventually turn into limestone or other geologic features, such as the White Cliffs of Dover.

The depth to which you dive with your class in investigating this question depends on the previous knowledge of your students about food webs.

OBJECTIVES

Although this question is a direct one, answerable in a few words, it opens the possibility for:
- Exploring the basic principles of food webs and energy flow.
- Pursuing a greater understanding of ocean food webs and later inquiry into human impacts on the sea soup and organisms dependent on a healthy ocean.

The National Science Education Content Standards addressed in these activities include:
- *Standard A:* Science as Inquiry
- *Standard B:* Physical Science—transfer of energy
- *Standard C:* Life Science—structure and function in living systems; populations and ecosystems

Background:

Animals spend most of their time finding food or avoiding becoming someone else's food. (Reproduction is their other major concern.) Therefore, "Who eats what?" is an important question for any habitat. In the *food chain* or energy

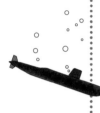

Diving Deeper: **Diving into Ecology**

*T*he word ecology is derived from the Greek word ecos meaning house or place to live. Ecology is the study of interactions between organisms and their environment including all abiotic (nonliving) and biotic (living or once living) factors.

An ecosystem includes the organisms in a certain area and their interactions with the physical environment. The energy flowing through the ecosystem will result in a diversity of species in different trophic or feeding levels and in a cycling of materials through the system. Remember, energy does not cycle, it flows from the ultimate energy source (usually the sun), through the living organisms, and is eventually lost as heat. Only about 1 percent of the light energy striking a plant is converted to plant matter. When a consumer eats a plant, only 10 percent of the energy from food consumption is converted to living matter for consumption by the next consumer level of the chain. Ninety percent is used for metabolism, moving, breathing, digesting, etc., or is lost as heat. Chemicals such as nitrogen, oxygen, and carbon cycle back through the environment. Ecosystems are linked by biological, chemical, and physical processes. The ecosystems discussed in the Sea Soup books are open ocean and coastal water ecosystems.

Every living thing exists as part of an ecosystem. No ecosystem is ever completely self-sustaining, because energy must enter the system to replace energy that has been degraded to a form in which it cannot do work. Usually the energy enters the system as sunlight energy, but in the unique hydrothermal vents and a few special terrestrial systems, simple chemical compounds such as hydrogen sulfide or methane supply chemical energy to initiate energy flow through the ecosystem. Phytoplankton play a critical role in supplying oxygen to the earth's ecosystem.

pyramid, energy is transferred from one organism to another. Usually the ultimate source of energy is the sun—sun or light energy that is converted to chemical energy during photosynthesis. Plants are called *producers. Consumers* (animals) that eat plants are called *herbivores*; animals that eat animals are called *carnivores*; animals that eat plants and animals are called *omnivores*.

Molecules that make up organisms are recycled back to the beginning of the food web by *decomposers* (usually bacteria and fungi) that break down the bodies and waste products, and release the nutrients and minerals back into the environment for use again by the producers.

A *food web* is a network of organisms in which one organism becomes food for another. Since many consumer organisms eat a variety of other organisms, food webs can become complex interactions between many organisms in one area.

Suggestions:

Before beginning the classroom activities, assign the question, "How many Phytoplankton does it take to fill a humpback whale?" from *Sea Soup: Phytoplankton*, then lead a discussion to ascertain what the children already know about food webs and energy flow in general. Use leading questions such as:

- How do animals get energy? (They obtain food and eat it. Digestion breaks the food into small pieces (molecules) that then pass into the blood and cells where complex processes release energy for use by the cell and whole organism.)
- Why do living things need energy? (Without energy, cells and organisms would die. Organisms use energy to make the building blocks of life, to repair damage, to move, to grow, to breathe, to do work, etc.)
- How do animals get energy from what they eat? (Food contains special chemicals that have energy in them. To get the energy out of the food, cells have a complicated process called cell respiration in which the food is gradually broken down into smaller and smaller parts. Each time it breaks down, energy is released from the food and trapped by a very special molecule called ATP. ATP carries energy to wherever energy is needed.)
- Do animals use all the energy contained in the food they eat? (No, a lot of energy is wasted as heat and undigested food. Only about 10 percent of the energy consumed is used by an organism.)

Activity 1: Who Eats Whom?

Objective:

Students construct a typical ocean food web using string and cards.

Background:

Nearly all life in the ocean is dependent on photosynthetic organisms. Only they and some very specialized organisms have the ability to manufacture their own food. With few exceptions, all animals are dependent on photosynthetic organisms. Plants produce. Animals consume. Producers form the first link in the food chain. The food chain represents the transfer of energy from one organism to another.

An example of an ocean food web includes: diatoms (phytoplankton) that are the basis of the food web; zooplankton such as copepods and crab larvae eat other zooplankton and phytoplankton; filter-feeding bivalves such as mussels and cockles filter zooplankton out of the water; codfish eat the bivalves; killer whales and humans eat codfish. In the ocean there are many individual food chains overlapping and intersecting to form complex food webs. Most marine creatures eat a variety of foods. This flexibility allows them to switch if one link in the chain is depleted.

Food webs usually end with a top predator on which nothing else preys. But, all living things die. Animals eat the dead organisms, and then bacteria and fungi attack and decompose the remains. Currents may carry the nutrients back to the surface where the phytoplankton use them in photosynthesis.

Materials:
- One "food web card" for each student
- String—long enough to make a loop to go over a student's head
- Index cards
- Marker
- Hole punch
- Yarn or string

Procedure:
1. Make enough "Food Web Cards" so each student will have one card, or make your own more relevant to your local ecosystem.
2. Organize your class into a circle.
3. Hand one card to each student to hang around his/her neck.
4. Give the end of the yarn or string to the "sun" person.
Discuss why the food web begins with the sun.
5. The students decide who or what organism depends on the sun.
Pass the yarn to one "producer" and back to the sun; then to another producer and back to the sun.
6. Then the yarn passes from herbivores back and forth to the producers. Each herbivore says something about what they eat and their role in the food web.
7. Then the yarn passes from herbivores to the predators on those herbivores. Again, each says something about their role in the food web.
8. Eventually the yarn passes to top predators who describe to the class how they are dependent on the other organisms.
When every student has had the yarn passed to them at least once, the game stops. Note how interconnected everyone is.
9. The teacher may introduce an environmental problem into the food web. What is the impact of a pollutant such as a plastic bag (eaten by the sea turtle instead of its food, a jellyfish)?
How does that compare with the impact of six-pack rings?
How does that compare with the impact of a fishnet that catches all the fish?
And, finally how does that compare with the impact of a pollutant such as an herbicide spill that kills all the phytoplankton?

A Typical
Ocean Food Web

SUN

Single-cell phytoplankton

Humpback whale

Krill

Crab larvae

Fish larvae

Fish

Herring

Copepods

Jellyfish

Puffin

Codfish

Mussels

Clams

Sea turtles

Lobsters

Eiders (sea birds)

Sharks

The teacher may introduce an environmental problem into the food web. What is the impact of a pollutant such as a plastic bag (eaten by the sea turtle instead of its food, a jellyfish)? How does that compare with the impact of six-pack rings? How does that compare with the impact of a fish net that catches all the fish? And, finally how does that compare with the impact of a pollutant such as an herbicide spill that kills all the phytoplankton?

Forty-mile-long drift net

Six-pack plastic rings

Plastic bag

Herbicide spill

Table 2

Food Web Cards

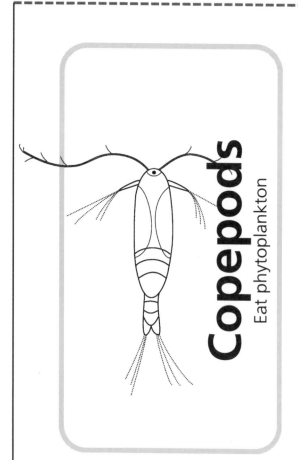

Copepods
Eat phytoplankton

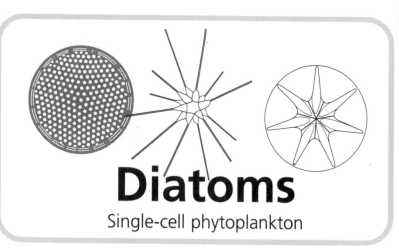

Diatoms
Single-cell phytoplankton

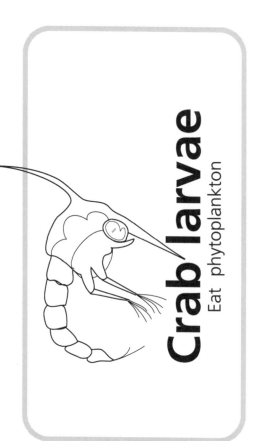

Crab larvae
Eat phytoplankton

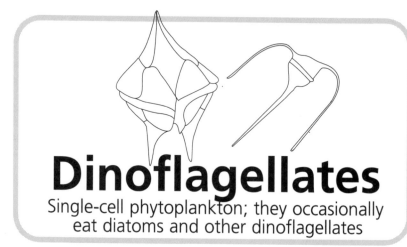

Dinoflagellates
Single-cell phytoplankton; they occasionally eat diatoms and other dinoflagellates

Fish larvae
Eat copepods, crab larvae, and phytoplankton

Mussels
Eat copepods, crab larvae, and phytoplankton

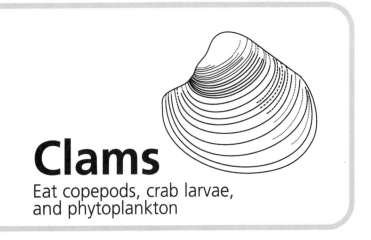

Clams
Eat copepods, crab larvae, and phytoplankton

Lobster
Eat dead mussels, clams, and fish

Herring
Eat krill and copepods

Sea turtles
Eat jellyfish

Forty-mile-long

Drift Net

Catches all the fish in the area

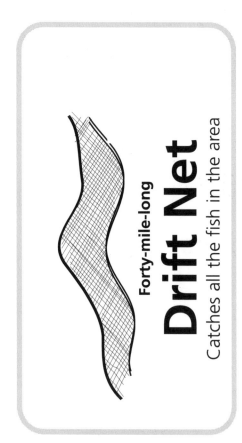

Six-pack

Plastic Rings

Choke pufin and eider

50

Jellyfish

Eat small fish and krill

TRASH

Plastic Bag

Eaten by sea turtle

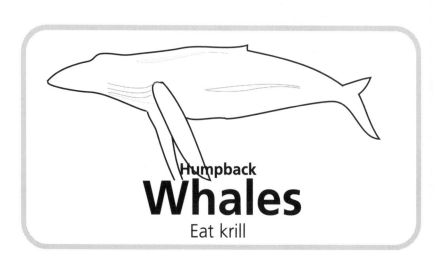

Humpback

Whales

Eat krill

Food Web Cards

Sun

Ultimate source of energy for all living things exept organisms living in deep ocean, hydrothermal vent or some thermal springs.

Eiders (sea bird)

Eats mussels and clams

Krill

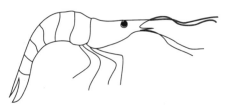

Eat copepods, crab larvae, and phytoplankton

Sharks

Eat small codfish

Puffins

Eat small herring

Diving Deeper: A Coral Reef Food Web

*B*esides the food web in Activity 1, we include an example of a coral reef food web for your information. Students can research information about the reef organisms.

Phytoplankton	Sea anemone
Algae (include non-planktonic algae that may grow on a reef or attached to shells and rocks)	Long-spined sea urchin
	Queen conch
Turtle grass	Shrimp
Moon jellyfish	Pufferfish
Green sea turtle, loggerhead sea turtle	Butterflyfish
Sponges	Squirrelfish or popeye anti
Sea walnut or comb-jelly	Octopus
Coral polyp	Squid or cuttlefish
Brittle starfish	Parrotfish
Sea horse	Spiny lobster
Sea fan	Grouper
Fan worm	Moray eel
Fireworm	Barracuda
Feather duster worm	Dolphin, porpoise
Coral crab	Nurse shark, ray

Activity 2: Make a Giant Food Web

If classroom space allows, the class can create a giant ocean-food web with creatures created out of papier-mâché, or large, two-dimensional cutouts hanging from the ceiling. The feeling created can be one of entering the sea soup in a tiny sub and observing a food web in action.

• Assign teams to create models (to scale if possible) of organisms that exist in particular roles in the food chain.

• One group can be charged with creating a variety of herbivores, small to large, that feed on phytoplankton. Another group creates the primary producers, the phytoplankton.

• Another group creates zooplankton including microscopic larvae to large jellyfish-like siphonophores—the Portuguese man-of-war. Remember zooplankton may be herbivores or carnivores.

• Another group creates larger, secondary consumers (carnivores).

• And, finally, a group creates the largest carnivores (whales, dolphins, seals, sharks, humans).

• Don't forget to create the sun to hang in the midst of the web! When the students complete their artwork, help them hang the critters around the room. You might choose to take a ball of yarn or string and, starting with the sun, cut lengths of yarn to attach the sun to the primary producers (phytoplankton), then the primary producers to the plant-eaters (herbivores), etc. The end result is a gloriously complex web of life.

Plankton Press

If you have not already done so, now is a good time to introduce the students to *Plankton Press*, the classroom newspaper about plankton. Explain that *Plankton Press* is an ongoing project for the duration of the Sea Soup unit. You may choose to dedicate a bulletin board as *Plankton Press* and post contributed articles, cartoons, editorials, and columns as they are assigned and completed. Or, you may expand the project to a computerized newsletter for your school, or the school's website. Or, you may ask students to develop a layout on the computer for the *Plankton Press* and actually print editions several times throughout the Sea Soup unit.

A contribution could include the following: Assign each student to do research and write an "interview" with a specific organism in an ocean food web.

Activity 3: How Do Tiny Zooplankters "Catch" Their Phytoplankter Prey?

Objective:

Students are challenged to think about how tiny animal drifters catch their prey.

Background:

Zooplankton are not just wanderers. Scientists have discovered that at least some of these drifters spend more than 80 percent of their time moving under their own power. Every night, when predators can't see them, zooplankters swim up towards the water's surface to feed on phytoplankton. As the sun rises, they sink into deeper water, sometimes traveling up to 300 feet each way. Using sidescan sonar, marine scientists track the diurnal movement of zooplankton from the ocean depths to the surface and back.

Some zooplankton, such as copepods and krill, remain drifters all their lives. Many other zooplankters spend only their earliest stages as drifters. Larval stages of fish, sea stars, barnacles, crabs, and lobsters look nothing like their parents at first. They grow by eating plankton and then change into forms that either settle to the bottom or develop the physical ability to swim.

Zooplankton are important to ocean food webs. We'll discuss a few common zooplankters.

Copepods are tiny cousins to lobsters and crabs; they are all in the class Crustacea and the phylum Arthropoda (animals with a hard exoskeleton and jointed appendages). They may be the most abundant animals in the ocean. Besides being eaten by larval fish, copepods are eaten by other copepods, krill, jellyfish, corals, and other small hydroids (tiny relatives of jellyfish and corals that grab copepods with their tentacles). The distinguishing feature of a copepod is its long antennae that sense the location of food and predators. A copepod swims by contracting its antennae against its body.

•*Krill* eat mostly phytoplankton, and almost all the creatures of the frigid Antarctic waters feed on one thing—krill (*Euphausia superba*). Krill look like pink shrimp, and they are about the size of a human thumb. Krill are eaten by penguins, seabirds, fish, squid, seals, and by some of the largest animals in the world, blue whales and humpback whales.

• *Larval animals* such as newly hatched fish, metamorphosing sea star and sea urchin larvae, larval crabs, lobsters, sea worms, clams, oysters, snails, and barnacles all feed on phytoplankton and zooplankton. For example, lobster larvae spend their first month after hatching floating at the ocean's surface feeding on plankton. They molt four times before sinking to the bottom and looking and acting like lobsters.

Materials:
   • Paper
   • Markers

Procedure:
Divide the class into groups of three or four children. Each group should select one person to take notes and be a spokesperson for the group.

Hand out one copy of Student Discovery Master 6: Design a Predator! to each group.

Extension Activity:
Students can research and write articles for the *Plankton Press* about the metamorphosis of fish, sea stars, urchins, worms, oysters, clams, and snails, and other marine organisms. The assignment allows for creative drawings of the larval stages. Students could also be encouraged to invent a marine critter and draw its stages of metamorphosis.

Student Discovery Master 6:
Design a Predator!

Name of scribe _____ **Date** _____

Class _____ **Names of group members** _____

Zooplankton may be tiny, microscopic animals or larger drifters. All zooplankton depend on phytoplankton as their source of food energy. However, to obtain the food, zooplankton must be adapted to locate the prey, get to the prey or get the prey to the predator, and actually catch and eat the prey.

Challenge:

Your challenge is to design a predatory zooplankter that catches and eats phytoplankton.

1. First, discuss how a zooplankter might prey on tiny phytoplankton. Brainstorm possible answers to these three questions. The group scribe lists the group's ideas on this paper. The discussion should take about ten minutes.

• How does the predator find the prey?

• How does the predator physically get to the prey?

• How does the predator catch the prey?

2. Then, the group brainstorms one best or favorite solution to each of the three questions.

3. Now, design the predator and draw it on paper. Each person in the group can draw their own version or the group may choose to illustrate their design with one drawing.

4. Label the drawing with a short explanation of which structure or mechanism is designed to address each question.

5. If class time allows, your teacher may ask each group to show and briefly discuss their designs.

6. Turn in this sheet and your designs. The scribe should make sure each student has signed his/her drawing.

Activity 4 **Where's the Energy Going? A Game!**

Objective:

This activity actively illustrates an ocean food chain and how the amount of phytoplankton limits the number of animals that can live in that ecosystem.

Materials:

- 30 phytoplankters (pieces of popcorn or 1-inch paper squares per "copepod")
- 1 small paper or plastic bag for each copepod
- 1 "herring" card for each student herring
- 1 colored armband for each copepod
- A different colored armband for each herring
- A different colored armband for each seal

Procedure:

1. Review the concepts of food web and energy flow with your class.

2. Mention that some copepods and other zooplankton migrate towards the ocean's surface at night to feed on phytoplankton. This may help the zooplankton avoid becoming prey instead of predators! Herring eat copepods wherever they find them. Seals usually hunt for herring during the day.

3. Move your class to a field or gymnasium, or rearrange your classroom so there is room for this activity.

4. Divide the class into three groups in the ratio of 1-seal: 4-herring: 20-copepods. (If you have more than 25 students add more copepods; if you have fewer than 25 students, decrease the herrings by one, and then decrease the copepods.) Two classes can easily be combined for this activity. Be sure to increase the amount of phytoplankton, too.

5. Hand each copepod an empty bag. This represents the copepod's "stomach" into which it will place "phytoplankton."

6. Instruct the class that you will be scattering phytoplankton around the area. Then, they should turn their backs to you while you distribute the phytoplankton around the large open space.

7. Ask the class to face you. Tell the copepods that on the count of "three," they should migrate to the phytoplankton area by jumping and moving their arms (since copepods swim by pulling movements with their antennae) to collect phytoplankton and place the food in their stomachs (the bags). At the end of 20 seconds the copepods stop collecting.

8. Now the herring are allowed to hunt for the copepods by swimming (hold two arms together in front and make fish-swimming movements). Copepods try to jump away from the herring. Depending on the size of the area in

which they are hunting, allow the herring to try to catch copepods until each has had time to catch one or more copepods (this may take as little as 10 seconds). Any copepod caught by a herring must give its bag of food to the herring and then the copepod leaves the ocean.

9. Any copepods still alive may continue to feed on phytoplankton. Herring may still try to catch copepods. But, now the top predator, the seal, appears on the scene and hunts herring. Any herring tagged surrenders its tag and bags to the seal and leaves the ocean.

10. Stop the game when one seal tags three herring.

11. Choose a few of the copepods and herring that had to leave the ocean to pick up and count all the remaining phytoplankton. Put the phytoplankton in one separate bag. Record the number of surviving phytoplankters.

12. Each surviving copepod should count the phytoplankton in his/her "stomach."

13. Each surviving herring should count the copepods in his/her "stomach."

Analysis:

1. To survive,
   - A seal must have at least 3 herring or it will starve.
   - A herring must have at least 4 copepods or it will starve.
   - A copepod must have at least 30 phytoplankton or it will starve.

2. Have the seals, herring, and copepods that did not catch enough food stand in three separate areas. (They did not survive.)

3. Have the herring and copepods that were food for their predators stand in two separate areas. (They did not survive.)

4. Record the numbers for use in discussion about the activity.
   - How many seals survived?
   - How many herring survived?
   - How many herring were eaten by the seal?
   - How many herring starved?
   - How many copepods survived?
   - How many copepods were eaten by herring?
   - How many copepods starved?
   - How many phytoplankters survived?

Discussion Questions:

1. What percentage of the original number of phytoplankters survived?

2. Is there enough phytoplankton to feed more copepods the next night when the copepods come to the surface to feed?

3. If not, what will happen to the food web?

4. Discuss what conclusions can be learned from this activity. (Food web has

a limited number of steps that can be supported by the amount of food energy in the system. It takes a lot of phytoplankton to support very few top predators. Too many predators may overeat their prey, and everyone loses.)

Extension Activities:
- Ask students to draw an imaginary scene from your miniature sub window. Include at least one kind of organism for each food level. Connect the food web by lines to show who eats whom.
- Louis Pasteur said, "The role of the infinitely small in nature is infinitely great." Ask students to write a short essay discussing whether they agree or disagree with the quote. They should include supporting evidence for their stand.

Activity 5: Investigating An Estuary

If you live near an estuary, where the river meets the sea, you may choose to do the following activity. Estuaries play a crucial role in introducing nutrients into the ocean and are important habitats for the reproduction and growth of commercially important species that feed on plankton as zooplanktonic larvae.

Objectives:
Students will:
- Research the characteristics and food webs of estuaries.
- Create an estuarine display or bulletin board emphasizing the importance of estuaries and plankton in coastal fisheries.

Background:
Most coastal towns have an estuary. It's where a river meets the sea. Because of the mixing of nutrient-rich fresh and salt water, estuaries are the most productive marine environments—acre for acre more productive than a cornfield in Kansas. The mixing of the nutritious waters of the ocean with the nutritious freshwater of the river creates a quiet, usually warmer environment ideal for the growth of phytoplankton and zooplankton. Many animal species use estuaries as a nursery for laying eggs and/or raising young, or as a place in which to rest and feed before continuing their migration. The biodiversity of an estuary usually exceeds that of the open ocean. The food web can be incredibly complex.

Whales, coral reefs, and sharks hold much allure, but there's a lot of action both good and bad happening in our estuaries. Without estuaries most of our marine resources would crash. Over 66 percent of commercially harvested marine organisms are dependent on healthy estuaries for at least some part of their lives. Perhaps they live in the open ocean, but their food comes from an estuary. Or, perhaps they live most of their lives in the open ocean, but they lay their eggs in an estuary.

This activity involves the students in researching information on estuarine ecology, particularly food webs. You may choose to supply information for them, or they may search the Internet or go to the library. The ultimate objective is for the students to develop an appreciation for the significance of our nation's estuaries.

The final product of this open-ended activity is to create a written report, bulletin board display, ceiling hanging display, or a large diorama illustrating estuarine food webs, with particular emphasis on the role of plankton.

Resources:

Two federally sponsored programs are committed to estuarine research, health, protection, and education.

1. Within NOAA (National Oceanic and Atmospheric Administration of the U.S. Department of Commerce), a unique group, the National Estuarine Research Reserve System (NERRS), provides information and education programs to the public about estuaries. About twenty-four estuarine research reserves exist around our coast. You may have one near you. Check their website listed at the end of this guide.

2. The Environmental Protection Administration (EPA) also has a section committed to monitoring and protecting estuaries. The National Estuary Program has informational brochures and websites appropriate for your students. An estuary near you may be included in this program.

Question 4
How Did Tiny Phytoplankton Build the Largest Structures on Earth?

From *Sea Soup: Phytoplankton*:

Coral reefs are underwater cities—busy, crowded, and colorful, just like cities on land. But they are even more densely populated, and often much larger than any human-built cities. The Great Barrier Reef along the eastern coast of Australia, for example, is over 1,000 miles long. Coral reefs are the result of a unique partnership between plants and animals: a type of phytoplankton called zooxanthellae and small, vase-shaped animals called coral polyps. The coral polyp takes calcium carbonate from the sea water to build itself a limestone house. But the coral polyps couldn't build a reef without the help of their plant companions, the zooxanthellae. Zooxanthellae are dinoflagellates that live inside a thin layer of tissue linking all the coral polyps together. (You can think of it like having small plants growing under your skin.)

Another amazing builder is a phytoplankter called the coccolithophore, a microscopic plant encased in 12 armored plates. As many as 1,500 plates could fit on the period at the end of this sentence. When millions of coccolithophores bloom at one time, astronauts in space can see the ocean's surface turn white.

After coccolithophores die, they sink to the bottom of the sea where they pile on top of each other until the weight of layer after layer of coccolithophores transform them into rock. After millions of years, the coccoliths, now hardened into limestone, are pushed up to the earth's surface to create towering structures like the White Cliffs of Dover in England or ones small enough to fit into your hand, such as a piece of chalk.

Objectives:

Students will:
- Learn about symbiosis (definition in Background).

- Discuss factors impacting coral reefs.
- Create a model of coral.

The National Science Education Content Standards addressed in these activities include:
- *Standard A:* Science as Inquiry
- *Standard C:* Life Science—characteristics of organisms; organisms and environments; regulation and behavior; structure and function in living systems; ecosystems
- *Standard F:* Science in Personal and Social Perspectives—changes in environments; resources and environment

Note to teachers using only Sea Soup: Zooplankton: *Corals eat zooplankton, and symbiosis is an important subject to investigate when studying ocean life. You may choose to assign students to research other symbiotic relationships such as those listed in the card game, "Find My Partner."*

Background:

Coral reefs are produced by billions of coral animals less than 1 centimeter (1/2 inch) wide. Each coral animal (polyp) makes its own hard skeletal cup. Many of these cups are cemented together to make up a colony. Reefs form when hundreds of hard coral colonies grow next to and on each other.

Although coral polyps are animals, they have a special dependency on photosynthetic cells. All reef-building corals have symbiotic photosynthetic cells called zooxanthellae living within their soft tissues. Up to 30,000 photosynthetic cells live in a cubic millimeter (less than 1/16 of an inch cubed) of living coral tissue. A close association between two different species in which neither species is usually harmed and, more likely, one or both species benefits from the relationship, is called *symbiosis*.

The zooxanthellae appear brown, golden, or brownish yellow. Most are dinoflagellates. They all live inside the bodies of the animals and most are found either in individual tissue cells or else in various body spaces between tissue layers. They grow and reproduce within the animal without being harmed by the animal.

Hard corals maximize their symbionts' exposure to sunlight. Shallow-water corals tend to be branched and the algae are piled closely throughout the soft coral tissue. Where light dims in deeper water, the corals tend to flatten out like tables to maximize the exposure of their algae to light. In dim light, the zooxanthellae have as much as four times the amount of chlorophyll per cell as those living in brighter light. At depths of about 60 meters, the light is too dim for zooxanthellae of reef-building coral colonies to survive.

Animals with zooxanthellae may change their form or behavior in order to maximize the zooxanthellae's exposure to sunlight. Anemones and corals with large numbers of zooxanthellae tend to have smaller tentacles since they depend less on capturing food and more on the food energy supplied by the photosynthetic cells within them. Corals with zooxanthellae keep their tentacles pulled in during the day to maximize the exposure of the zooxanthellae to light. At night, the tentacles extend and the coral feeds on plankton.

Diving Deeper: Coral Polyps

Each coral animal (polyp) removes carbonate and calcium from seawater and deposits it as a hard calcium carbonate skeleton. Each polyp sits in a tiny cup. Many of these cups are cemented together to make up a colony. Reefs form when hundreds of hard coral colonies grow next to and on each other. New polyps build their skeleton on the old at a rate of 6–12 inches a year in some reefs and as slow as a quarter of an inch per year in other corals. This results in the build-up of corals into huge limestone reefs. The Florida Keys and some Caribbean islands are ancient reefs left out of the water as sea level changed.

Diving Deeper: Details of a Deep Relationship

One of the most striking features of marine organisms is the large number of close, special relationships that can be observed among many different species. Not predator-prey relationships but close associations between two different species in which neither species is harmed and, more likely, one or both species benefits from the relationship. This is called symbiosis.

Many types of photosynthetic, single-celled organisms have symbiotic relationships with a wide variety of marine organisms. Photosynthetic cells living in an animal are called zooxanthellae. Because the cells require light, symbiotic relationships with them are restricted to shallow waters where sufficient light is present.

Zooxanthellae and coral animals are so intimately connected that they behave as a single organism. The coral polyps generate carbon dioxide and ammonia, wastes for the coral animals but perfect fertilizers for zooxanthellae which absorb the compounds directly from the animal tissue. This system is so efficient that some coral species release virtually no nitrogen into the surrounding water; their photosynthesizing symbionts absorb ammonia continuously. The photosynthetic cells generate oxygen as they photosynthesize. Most of this oxygen is absorbed directly by the animal tissues. The zooxanthellae also make sugars, amino acids and other organic compounds that are absorbed directly by the coral tissues. Somehow, zooxanthellae help hard corals manufacture and deposit their calcareous skeletons. Deposition of calcium carbonate may be directly stimulated by organic molecules given off by the algae or possibly the algae's removal of coral waste products may remove the inhibitory effects of the waste products on skeleton production.

Dinoflagellates change form when they live as symbionts. They lose their flagellae and characteristic grooves around the body. The cell walls are much thinner. If the zooxanthellae are removed from the host and grown outside the animal in culture, they develop the characteristic flagellae, cell walls, and other structures typical of a free-living form.

Soft corals (those that do not make a hard, calcareous skeleton) often lack zooxanthellae. They obtain nutrients by ingesting plankton. Because they do not depend on photosynthetic symbionts, many soft corals live in dimly lit caves, on overhanging ledges, and on the deeper sea bottom.

Diving Deeper: Other Organisms with Zooxanthellae

Other organisms have symbiotic relationships with dinoflagellates. Single-celled zooplankters related to amoebas and parameciums may have zooxanthellae living within them. Many marine sponges (which are multicellular, but simple animals) have zooxanthellae living in them. Since tropical waters are typically low in nutrients, the ability of the animals and zooxanthellae to help each other obtain nutrients and energy-containing molecules helps both partners to survive.

Within the group that includes anemones, jellyfish, and corals, symbiosis is extremely common. The upside-down jellyfish (Cassiopeia xamachana) lives upside down in quiet, shallow water soaking up the Florida sunshine, allowing its zooxanthellae to photosynthesize. It can capture its own food, but in good light it can exist on the food products produced by the zooxanthellae. Most tropical, shallow-water anemones, soft corals, sea fans, whips, and stony corals have symbiotic zooxanthellae within their tissues.

Giant clams are Molluscs, the group that includes snails, clams, and octopus. Giant clams have zooxanthellae living within their tissues. Giant clams live in the tropical waters of the Indo-Pacific where they may reach 4 feet long and weigh 584 pounds. Unlike most clams, giant clams lie on the bottom or bore into corals in a position with the opening between the shells facing up towards the light. As they gape the shells, the extensive, brightly colored tissues containing the zooxanthellae are exposed to the sunlight. An unusual relationship exists between giant clams and their zooxanthellae. When the zooxanthellae cells are old and degenerating, the clam has special cells that can distinguish healthy cells from unhealthy cells, and it consumes the dying cells. This ability enables the clam to use the nutrients contained within the old zooxanthellae cells rather than losing the nutrients when the cell dies.

Activity 1: Just the Two of Us

Objective:

Students play a card game similar to "Go Fish" in which they try to match symbiotic pairs. This activity emphasizes the wide variety of symbiotic relationships that exist in nature.

Materials:

- Stiff paper or cardboard for making cards
- Markers
- Glue
- Scissors

Procedure:

1. Before class, copy the "Find My Partner" cards on heavy paper. Cut the pages into cards. Make one complete set for every group of four students.
2. Ask the class to read about, "How did tiny phytoplankton build the largest structures on earth?" in *Sea Soup: Phytoplankton*. For classes using only *Sea Soup: Zooplankton*, there may not be an appropriate reading in the book, but symbiosis is an important subject to investigate when studying ocean life.
3. Discuss the reading and symbiosis (using background information).

4. Divide the class into groups of three or four students.

5. Hand one set of cards to each group.

6. Explain that this game is played like "Go Fish." The object is to collect pairs and be the first one to "go out."

Note: When a student has a "match," he/she must read the information on the match pair to the group. One student shuffles the cards and deals five cards to each person. Then, he/she places the remaining card in the middle. Any one with a matched pair reads them out loud and places them in front of him/her. Then, student "A" to the left of the dealer asks the person to his/her left (student "B") for a card to match one already held. If student "B" does not have a matching card, then student "A" is told, "Find my partner" and must take a card from the pile. If student "B" has a match, he/she must give it to "A" who then reads the cards and places the matched pair down. Student "A" gets to ask another person for a card. If the person does not have a matching card, then "A" takes a "Find My Partner" card from the pile. Play continues around the circle until one student is out of cards. That person "wins."

Algae
Upside-down Jellyfish

You give me a safe place to live.

Upside-down Jellyfish
Algae

Find My Partner!
Symbiotic Partnerships in Nature

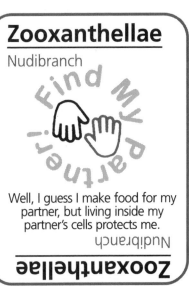

Zooxanthellae
Nudibranch

Well, I guess I make food for my partner, but living inside my partner's cells protects me.

Nudibranch
Zooxanthellae

Whale
Barnacle

What do I get out of having barnacles attached to my body? I don't know!

Barnacle
Whale

Reef Fish
Pacific Wrasse (fish)

Little crustaceans attach to our skin and bother us. I visit this partner to get rid of the parasities.

Pacific Wrasse (fish)
Reef Fish

Buffalo
Cowbird

The little pesky bird eats bugs off me. That's cool!

Cowbird
Buffalo

Clownfish
Anemone

I make a thick layer of mucous to mask the chemical stimuli that help fire my partner's stinging cells

Anemone
Clownfish

Honey Bee
Flower

I gather a flower's sweet nectar to feed the young in the hive.

Flower
Honey Bee

Goby
Grouper

I'm small but mighty. Big fish come to me for a cleaning and I get to eat their parasites.

Grouper
Goby

Shrimp
Fish

I'm a special kind of shrimp that cleans dead cells and parasites off fish.

Fish
Shrimp

Remora Fish
Shark

I attach to sharks to get a free ride and eat the scraps dropped by the shark.

Shark
Remora Fish

Zooxanthellae
Coral Polyp

Thousands of us live inside the coral. We get protection and make food for the coral.

Coral Polyp
Zooxanthellae

Hermit Crab
Sea Anemone

My partner gives me camouflage from my predators (they can't see me).

Sea Anemone
Hermit Crab

Bottom-dwelling
Shrimp
Blenny

You dig the burrow and I'll watch for predators.

Blenny
Shrimp
Bottom-dwelling

Cardinalfish
Long-spined Sea Urchin

I seek safety among your long spines.

Long-spined Sea Urchin
Cardinalfish

Shrimp
Spanish Dancer

With very long legs, my leggy mate and I seek protection by nestling by your gills.

Shrimp — Spanish Dancer

Giant Clam
Algae

You make food for me.

Giant Clam — Algae

Upside-down Jellyfish
Algae

You photosynthesize and make food for me.

Upside-down Jellyfish — Algae

Pacific Wrasse (fish)
Reef Fish

We set up "cleaning stations" where we clean and eat parasites off visitors at the reef.

Pacific Wrasse (fish) — Reef Fish

Nudibranch
Zooxanthellae

I'm like a shell-less snail with many fingerlike projections. My partner makes food for me.

Nudibranch — Zooxanthellae

Barnacle
Whale

I get a free ride on my partner. Plankton food comes to me as my partner swims.

Barnacle — Whale

Cowbird
Buffalo

I follow this big animal around and eat the insects it kicks up and that live in its fur.

Cowbird — Buffalo

Anemone
Clownfish

My stinging tentacles do not seem to bother this colorful animal nestled among my tentacles.

Anemone — Clownfish

Flower
Honey Bee

My partner carries my pollen to fertilize another flower.

Flower — Honey Bee

Grouper
Goby

I'm a huge fish but I let a small goby fish pick parasites off my gills and mouth.

Goby
Grouper

Fish
Shrimp

I wait in line to visit the cleaning station set up by a shrimp.

Shrimp
Fish

Shark
Remora Fish

I let the remora travel on me. It's no trouble.

Remora Fish
Shark

Coral Polyp
Zooxanthellae

I can't live without my partners; they help me make my food.

Zooxanthellae
Coral Polyp

Sea Anemone
Hermit Crab

If you tickle me with your claws, I'll abandon my rocky perch and attach to the top of your home.

Hermit Crab
Sea Anemone

Blenny
Bottom-dwelling shrimp

At the first sign of danger, we both will dive for cover.

Bottom-dwelling shrimp
Blenny

Long-spined
Sea Urchin
Cardinalfish

I don't think you help me, but you don't hurt me.

Cardinalfish
Sea Urchin
Long-spined

Spanish Dancer (fish)
Shrimp

Your presence doesn't bother me.

Shrimp
Spanish Dancer (fish)

Algae
Giant Clam

You give me a safe place to live.

Giant Clam
Algae

Extension Activity:

Two interesting questions which advanced students might enjoy researching or discussing are:

• How do the symbiotic zooxanthellae get passed to the next generation of animal? (Since most marine animals reproduce by releasing their egg and sperm cells into the ocean water, how do the fertilized eggs and developing animals end up with zooxanthellae in their tissues? Either the zooxanthellae cells must be passed on directly from the parents to the eggs or larvae (sperm cells are too small to include zooxanthellae), or each developing young animal must acquire zooxanthellae cells from its surrounding environment. Among some corals, zooxanthellae do not appear in the young animal until a week or two after the larvae have settled. The method used by giant clams is not known. Some evidence indicates that algae are transmitted through eggs. Another possibility is the clams receive them from the surrounding corals. Sponges ingest the free-living dinoflagellates from the surrounding environment. Some animals may release a chemical that attracts the zooxanthellae to the developing young. Some animals might eat another animal that contains the symbiont.)

• Since corals exist in huge numbers on reefs, and the water surrounding the reefs is extremely poor in plankton, how is it possible for the coral animals to obtain enough food? (Studies show that a large part of the plankton population is not from the open ocean but living in the reef itself. These plankters stay on the bottom during the day, and rise into the water over the reef at dusk. Corals open up their tentacles at night to feed on the plankton. During the day, the coral close up. Their zooxanthellae are exposed to sunlight and photosynthesize. Experiments in which the zooxanthellae are removed or the coral is deprived of light show that some corals can survive by feeding on plankton at night. However, they do not grow. Corals that feed at night and whose zooxanthellae photosynthesize during the day gain weight and grow. Obviously, zooxanthellae are furnishing food energy to the corals. Zooxanthellae significantly increase the calcification rate of corals and therefore the rate of growth of the coral colonies. Light enhances oxygen production that in turn stimulates coral metabolism leading to increased calcium carbonate deposition. Rapid calcification is needed to maintain the reef against destructive forces such as storms, anchors, and parrotfish.)

Activity 2: Build a Reef

Objective:

Students create coral polyp puppets and create a coral reef.

If you have not already done so, discuss the behavior of coral polyps during the day and night (see sidebar).

Materials:
- Pictures of coral polyps and coral colonies
- Template
- Oak tag or heavy paper
- Tape
- Scissors
- Thin plastic gloves
- Markers—yellow and green
- Old bed sheets or curtains
- Goldfish crackers or baby carrots

Procedure:

A. Make a Coral Polyp.

1. Cut several templates for the volcano-shaped cone (hard coral cup).

2. Make one hard coral cup as a demonstration.

3. Cut many X's about 4 or 5 inches long in some old bedsheets. Each sheet represents one colony of corals.

4. Suspend the sheets between chairs with room underneath the sheets for a group of students. Keep all the sheets close together.

5. Organize the materials.

6. Review the photographs of corals with your class.

7. Each student traces the template onto oak tag and cuts it out.

8. Tape the oak tag into a volcano shape with the hole large enough for a child's hand to poke through the top.

9. Have each child use a marker to make dots, "zooxanthellae," (either green or yellow) on the *back* of the fingers and *back* of the hand of a plastic glove. The glove now represents a single coral animal (polyp) with zooxanthellae. When a child extends the fingers of the glove out straight, it represents coral tentacles feeding at night. When the fingers (tentacles) are curled inward and exposing the zooxanthellae, it represents a coral polyp during the day.

10. Have each child put their gloved hand through the cone and wriggle their fingers like the feeding tentacles of the polyp.

B. Make a Coral Colony.

1. Have groups of children gather under a sheet, one at each opening, with their coral polyp puppets. Explain that they are independent animals, but they are all connected in a unit or colony. A colony of coral may contain thousands of coral polyps.

2. Each group chooses what kind of coral it is.

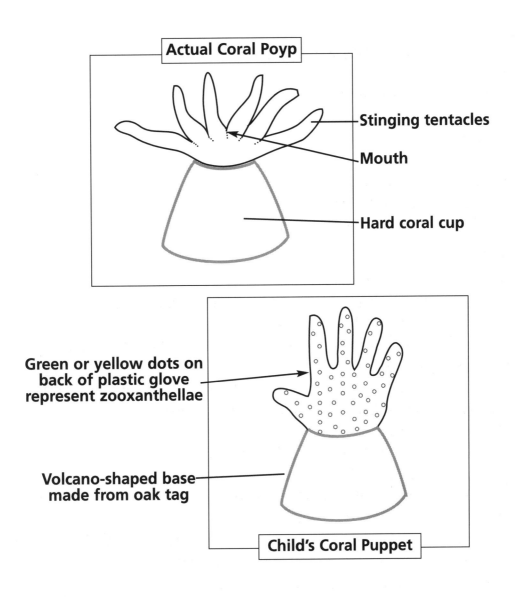

Actual Coral Poyp

Stinging tentacles

Mouth

Hard coral cup

Green or yellow dots on back of plastic glove represent zooxanthellae

Volcano-shaped base made from oak tag

Child's Coral Puppet

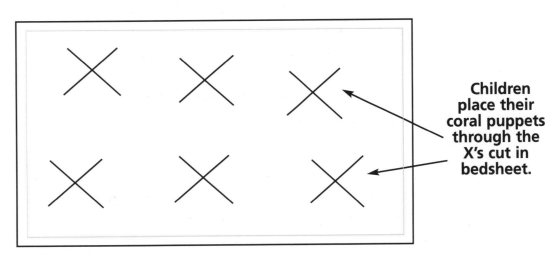

Children place their coral puppets through the X's cut in bedsheet.

Coral Puppet Template

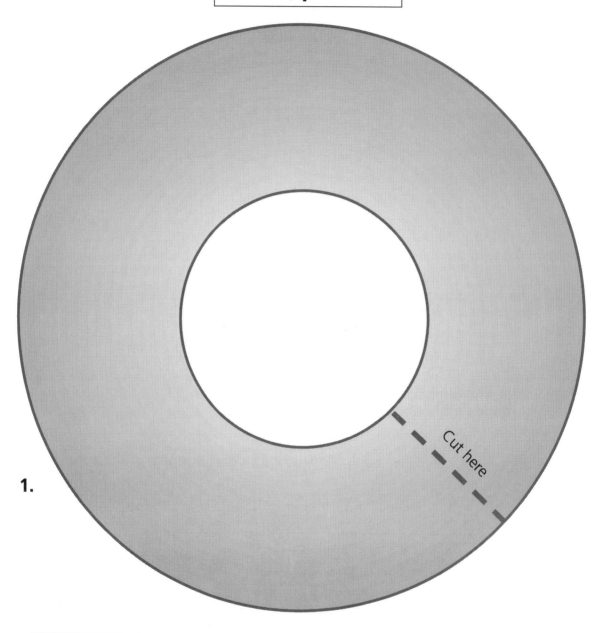

Cut here

1.

1. Use template to cut out one circle for each coral puppet.
2. Adjust puppet into a cuff shape around each student's wrists with tape.

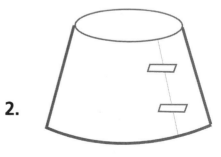

2.

3. Practice nightime, predatory, feeding behavior. Now ask the "coral polyps" to imitate daytime corals.

4. To demonstrate feeding behavior, pass out "plankton" (goldfish crackers) when the students reach up through the sheet with their gloved hands. Like polyps, the students need to pull their "tentacles" in to eat.

5. At night another interesting thing happens on the reef. Neighboring corals of different kinds go to war! Because space is limited on a reef, corals will actively extend their tentacles and "sting" neighboring polyps of a different kind of coral when the coral gets too close. White scars on dead coral mark where the attacks took place.

6. To demonstrate "coral wars," have the outermost polyps of each colony reach out with their tentacles and gently tap a neighboring coral polyp of a different kind.

7. Ask: Can colonies move away from each other? (No, because corals have hard cups, they cannot move.) In a crowded reef, how do corals get more space? (They have to eliminate their neighbors.)

The preceding activity was modified with permission from *Reading the Environment: Children's Literature in the Science Classroom* by Mary M. Cerullo (Portsmouth, NH: Heinemann, 1997).

For review or homework:

Using the model just constructed, ask the groups to discuss how the coral gets its energy.

• Does it use its tentacles during the night, or during the day? How do the zooxanthellae help the coral?

• When do the zooxanthellae make their food?

• What does the coral do to help the zooxanthellae make their food?

• What does the coral get out of all this? (What is the advantage to a coral of having zooxanthellae?)

• Any ideas on why a coral wouldn't be better off keeping its tentacles out all the time so it could feed twenty-four hours a day? (Think about predation by animals on the coral polyps.)

Diving Deeper: Corals Look Different at Night!

Corals with zooxanthellae keep their tentacles tucked in during the day. This exposes the zooxanthellae to sunlight for photosynthesis. As the sun goes down, the long, finger-like tentacles come out and grab and eat plankton and other organisms such as small fish. Hard corals are predators at night, and, with the help of zooxanthellae, are photosynthesizers during the day!

Extension Activity:

The following poem appeared in an 1872 collection of verses, *Sing Song,* by Christina Rossetti.

> O sailor, come ashore,
> What have you brought for me?
> Red coral, white coral,
> Coral from the sea.
>
> I did not dig it from the ground,
> Nor pluck it from a tree:
> Feeble insects made it
> In the stormy sea.

You may choose to discuss the poem in class or assign the following questions as written homework.

• Who is speaking in each of the lines?

• What do you imagine the personal situation was for each voice?

There are two things in the verses that would not be true or appropriate today. What are they?

(Answer:

1. In many countries today it is illegal to remove any live or dead coral from the seas or beaches.

2. Insects do not make coral; the coral skeleton is secreted by living polyps in the phylum Cnidaria, class Hydrozoa—soft bodied animals like anemones and jellyfish—not the phylum Arthropoda, class Insecta!)

• What is the impact of stormy seas on live coral?

(Answer: Ocean waves do little harm to corals since the corals grow beneath the water's surface. However, severe storms may create currents that carry sand onto the corals and smother them. Sudden freshwater runoff may change the water's salinity and harm the coral polyps.)

Activity 3: What Happens If...?

Objective:

Students will research and discuss the threats to the world's coral reefs.

This activity is most appropriate for fifth, sixth, and seventh grades. Teachers of younger children may choose to spend some class time discussing the topics.

Background:

The following information serves as a starting point for you and your students. You may choose to copy it for students or use the information yourself in leading a discussion of the issues. Websites may be helpful to the students if you choose to have them do research into the topic and contribute to the *Plankton Press.*

Corals in Hot Water

Coral reefs grow best in warm, relatively shallow water, so they are found only in a wide band on either side of the Equator. However, if the water gets too hot it can be a disaster for them. Divers have watched in dismay as whole reefs turned ghostly white in a matter of days. Scientists call this coral bleaching, a natural disaster in which some or all of the zooxanthellae leave. If ocean conditions return to normal, the zooxanthellae that stayed behind begin to grow again and the coral will recover. If not, the corals die.

Although localized cases of coral bleaching had occurred earlier, starting about 1980 scientists began to observe cases of mass bleaching where huge areas of coral reefs—sometimes thousands of miles across—suffered. Some of these areas were far from land, people, and obvious causes.

Why does coral bleaching take place? That is the mystery that Thomas Goreau of the Global Coral Reef Alliance has been investigating for nearly twenty years. In many places, he has found that up to 80 percent of the corals are affected. He says, "When the coral loses the algae, it's not only lost its color and most of its food supply, it completely stops growing and reproducing. If the stress continues or gets worse, bleaching is just the first step towards dying." Tom strongly believes that coral bleaching is caused by worldwide global warming. "The 1980s and '90s have been the hottest decades on record since temperature measurements began 150 years ago. "When we get just one degree above normal for one month, most of the corals will recover. If it is one-and-a-half degrees, maybe half will die. When it's two to three degrees, they almost all die. We are seeing a number of species disappearing in some areas." A possible cause of global temperature rise is the increase in carbon dioxide from the burning of trees and fossil fuels increasing the greenhouse effect. (See Question 7.)

From *Ocean Detectives: Solving the Mysteries of the Sea.* Mary M. Cerullo (Austin, TX: Raintree Steck-Vaughn Publishers, 2000).

Threats to Our Coral Reefs

Our coral reefs are declining precipitously. Coral reefs support almost a million species including 25 percent of all fish species, they protect coasts from erosion and storm surges, provide sources of food and medicines, act as recreation areas, and provide nursery areas for the young of many species. But, more than two-thirds of the earth's coral reefs are dying. Ten percent may be beyond recovery and 30 percent are in critical condition and may die by 2030. The causes include dynamite and cyanide fishing, chemical pollution, fertilizers, sewage, siltation from sediments carried by freshwater runoff from land, and climate change. Pollution that increases the water's cloudiness blocks the sunlight from the zooxanthellae. Deforestation (extensive logging) results in soil erosion into rivers, increasing cloudiness and nutrient content of water. Sediments smother the polyps. Chemical fertilizer runoff and sewage discharge adds too many nutrients; other algae may grow and cover the coral polyps. Heavy metals, pesticides, and oil kill reef organisms.

Power and desalination plants kill plankton as water is drawn through filters into their machinery. Hot water discharged from plants is lethal to organisms used to the stable temperatures of tropical seas. Sea level changes may make shallow water too warm for polyps. They then eject their zooxanthellae, resulting in bleaching.

Physical damage to reefs by recreational fishermen and boaters carelessly anchoring, tossing garbage, and running aground on coral compound problems caused by divers and snorkelers who stand on coral and kill the polyps, break the corals, or stir up sediments with their fins.

Other Factors Impacting Coral Reefs

Storms carry sand onto the reefs. Small amounts of sand may wash off the reefs by currents and waves. However, large deposits of sand block sunlight from reaching the zooxanthellae and prevent the coral polyps from grabbing food organisms with their tentacles. The reefs die.

Corals may seem safe inside their deep, hard skeletons, but they are prey for many predators. Urchins, worms, snails, mussels, sponges, sea stars, and fish eat the soft coral animal and bore into the reef. Parrotfish dine on corals. When snorkeling or SCUBA diving you can hear quiet crunching. Usually you'll see a rainbow-colored, foot-long fish biting mouthfuls of coral, chewing, and, if you watch closely you may see the fish excrete or spit out white sand—chewed coral. Some of the sand of the white beaches of the Caribbean has passed through the mouth of a parrotfish!

The difference between storm damage, parrotfish, and coral bleaching is that coral bleaching may be caused indirectly by human activity that may be causing global warming. Reefs are adapted to surviving parrotfish and storms, but not drastic changes in environmental temperature.

Procedure:
    1. Briefly discuss factors impacting coral reefs.
    2. Assign small groups to research one factor impacting reefs. Allow sufficient

class time for their research on the Internet or for using library and other resources.

3. You may choose to assess their work by asking for written or oral reports, bulletin board displays, or contributions to *Plankton Press.*

Moderate a class discussion on"What happens if...?

• If humans impact the reefs.

• If the reefs are destroyed.

• What are the different ways humans might impact the reefs?

• Are reefs healthy?

• If not, what is hurting them? What are the causes of reef destruction? Where is it occurring?

• What is the impact of reef destruction on the coastal lands currently protected from ocean storm surges and waves by the reefs? On the diversity of organisms that live in and around reefs? (What animals depend on reefs for food, protection, reproduction?) On health of the oceans? On humans? (Some people are dependent on fisheries, tourism.)

Extension Activity:

You may choose to use the above background information supplemented by updated information from the Internet (*New York Times,* NOAA, EPA) to have the students write contributions to the *Plankton Press* about the "State of the Reefs."

Question 5
Why Is the Ocean Blue (or Not)?

The apparent color of the ocean is influenced by the wavelengths of light absorbed and reflected by the water and any matter in the water. Blue water is clear water in which little, if any, phytoplankton grow. Nothing reflects the light back into space. Green water appears green because the phytoplankton reflect green wavelengths of light back into space. Where there's green, there's life! Zooplankton graze on the abundance of food. The ocean may also appear white if there is a bloom of coccolithophores, red if certain dinoflagellates are blooming, or glow at night when phosphorescent dinoflagellates bioluminesce.

The view from an airplane reveals areas of water rich in plankton and areas almost devoid of life. Estuarine areas where freshwater rivers loaded with nutrients mix with sea water are abundant with plankton. A similar phenomenon exists where nutrient-rich water from the deep ocean canyons wells up towards the surface due to the topography of the ocean's bottom. When deepwater currents encounter geologic structures that force the water up, the nutrient-rich water mixes with surface waters and phytoplankton abound. This is called an *upwelling*.

Where there are phytoplankters, there are zooplankters. Areas rich in phytoplankton support a multitude of ocean life and attract commercial fishing vessels. Using advanced satellite technology, scientists and fisheries biologists can measure the productivity of the waters and predict the impact on the fisheries.

The National Science Education Content Standards addressed in these activities include:

Standard A: Science as Inquiry

Standard B: Physical Science—heat; motion

Standard C: Life Science—structure and function

Standard D: Earth and Space Science—structure of the earth system

Standard E: Science and Technology—abilities of technological design; understandings about science and technology

Activity 1: Analysis of NASA Satellite Images

Objectives:

Students will:

• Recognize that there are infrared and normal images taken by satellite.

• Recognize the differences in ocean surface water temperatures using satellite images and predict where coastal upwellings occur.

This activity may be more appropriate as written for grades five, six, and seven. Teachers of elementary grades may choose to cover the topic by having students create a mural of the Caribbean and a more nutrient-rich water body such as the Gulf of Maine or Puget Sound. The teacher could download color copies of satellite images or copy from a resource ahead of time. Students could do some research and then write a short essay on differences between the two areas.

Background:

In recent years satellite images of the earth have yielded amazing information about the state of our oceans. Infrared satellite images show surface temperature patterns. These give clues about the dynamics that play a major role in the distribution and abundance of nutrients and marine life. Ocean surface temperature data allows prediction of weather patterns and upwellings of cold, deep-ocean waters.

When winds or currents move surface water away, it is replaced by colder, nutrient-rich water that moves up from below. This phenomenon is called upwelling. The cold deep-ocean water carries large quantities of nutrients used by phytoplankton. Since phytoplankton form the base of the ocean food web, where there are nutrients, there will be phytoplankton; where there are phytoplankton there will be a diverse assemblage of prey species including hoards of zooplankton. This is of particular importance to commercial marine fisheries.

Satellite images similar to common, color photography also show real color differences in the sea soup. *Sea Soup: Phytoplankton* shows a wonderful satellite image of a coccolithophore bloom and another of phytoplankton abundance. Other images available on the Internet or in books show red tides and healthy phytoplankton populations.

In this activity students will access satellite images of ocean temperatures and images of differences in phytoplankton abundance.

Diving Deeper: **Upwellings**

*T*he coast of Peru normally experiences a large upwelling of cold water that ultimately results in large anchovy catches. If the trade winds weaken or reverse direction, upwelling decreases, water temperature increases, nutrients decrease, and fewer plant and animals survive and reproduce. In abnormal years, when El Nino occurs, no upwelling occurs. The phytoplankton "crop" fails and the anchovy industry slumps.

On the East Coast of North America, the continental shelf precipitously drops to the abyss and sets up ideal conditions for upwelling. Georges Bank in the Gulf of Maine perfectly illustrates the interaction between the physical formation of the ocean bottom and the biodiversity of the food web.

In colder climates such as the Gulf of St. Lawrence in Quebec, cold ambient temperatures cause surface water to sink and to be replaced by the nutrient-rich deep water. Huge phytoplankton populations support huge zooplankton populations. A wide variety of baleen whales including finbacks, humpbacks, minkes, sei, right whales, and an occasional blue whale feed in this area.

Diving Deeper: **Ocean Temperatures and Satellite Imagery**

*T*he Landsat satellite collects information in many bands of the electromagnetic spectrum including visible light (red, green, and blue). During the day, two data streams—one visible and one infrared image—are transmitted. At night, both channels are infrared. Unlike a map, a satellite image is an array of cells aligned in rows and columns. Each cell has a value based on the reflectance characteristics of that part of the earth's surface that it represents. A cell representing a dark road surface will have a relatively low value; a cell composed entirely of snow-covered mountains will contain a very high value in the visible spectrum. Deep water generally appears black because it reflects little solar radiation. But, in the thermal infrared band, ocean water contains a variety of values based on relative surface temperature. Grouping the thermal data from cells taken above the oceans gives them unique colors and patterns emerge.

Diving Deeper: **Real Color Images and Satellite Imagery**

*R*eal color differences in the sea soup show on satellite images similar to those from a conventional camera. Coccolithophores are tiny cells covered by platelets known as coccoliths—round, ornate structures made of calcium carbonate (chalk). Coccolithophores are always present in some coastal waters, but sometimes a bloom occurs in which coccolithophores are prevalent. During these blooms, which may last up to a month, the cells grow and divide twice a day. They produce and continually shed their coccoliths. Dozens of coccoliths from each cell color the water white and the bloom appears as a white area in the satellite image. Such blooms may cover 15,000 square miles. Satellite imagery tracks temperature changes and blooms. Blooms of nontoxic algae may be followed by increases in organisms that feed on the phytoplankton. Tracking blooms may help commercial fisheries find their prey.

Materials:

To obtain infrared satellite images of ocean water temperatures and real color images showing relative phytoplankton abundance access one of the websites listed in the resource section at the back of this guide under Websites: Satellite Images of Phytoplankton.

Diving Deeper: The Gulf of Maine and the Caribbean Sea

The Gulf of Maine is a "sea beside a sea" bordered by Massachusetts, New Hampshire, Maine, New Brunswick, and Nova Scotia, and separated from the Atlantic Ocean by shallow, underwater areas of sand and gravel. Fresh water flowing into the Gulf from many rivers makes it colder and less salty than the outer Atlantic. A gyre, an ocean current flowing counterclockwise in the Gulf, circulates nutrients, food, and pollution. Nutrients are washed from the land and carried by rivers, or are churned up from the ocean floor. They support vast quantities of phytoplankton, zooplankton, and huge schools of fish. The strong currents keep mixing nutrients into the waters. The shallow areas are perfect for photosynthesis. Seasonal temperatures vary along with the amount of nutrient mixing. Therefore, phytoplankton numbers vary, too.

The tropical waters of the Caribbean Sea are warm, clear, shallow, and very salty. Located southeast of the Gulf of Mexico, the Caribbean Sea is about 500 miles wide by about 1,200 miles long with the Caribbean Current flowing from east to west. The temperature, depth, and chemical and physical characteristics are very similar throughout the whole area. The area is devoid of phytoplankton compared to the Gulf of Maine. No seasonal variation in phytoplankton levels or sea surface temperatures occur.

Procedure:

1. Read "Why is the ocean blue (or not)?" from *Sea Soup: Phytoplankton,* or another appropriate reading.

2. Ask students to make a list of differences between the Caribbean and the Gulf of Maine or another nutrient-rich body of water such as Puget Sound. You may choose to make this a research project or you may list what facts the students know, and then supply them with the information (see sidebar).

3. Using images showing differences in sea surface temperature, ask the students to determine the scale and how the differences in temperature are indicated. For example, do they think that darker water is warmer or colder? Does the scale show temperature differences in color?

4. Once they have determined how the image depicts colder water, locate the colder seas. In general, where is the warmest water in the image? Why is it warmer? Have them look for areas of cold water along the coast that are caused by upwelling.

5. Find examples of warm-water and cold-water currents, and ask them to try to find the names of these currents. Examples in the North Atlantic include the Gulf Stream and the Labrador Current.

6. With an image that depicts phytoplankton density, use the color scale on the computer screen to determine where the areas of high phytoplankton density are located.

7. In general, where are the areas with higher phytoplankton levels? How does this correlate with the information that they learned from the sea surface temperature image?

8. Based on information they have from the satellite images, ask them to predict where they would find major fisheries of the world (where there are cold-water currents or cold-water upwellings).

9. You may choose to have your students research the locations of major fisheries of the world. They could then compare their research findings with their predictions from the satellite images.

The preceding activity was modified with permission from the curriculum *Space Available: Aquatic Applications of Satellite Imagery*, published in 1997 by the Gulf of Maine Aquarium, Portland, ME.

Extension Activity (for sixth and seventh grades):
Objectives:
- Use bathyometric maps to predict where upwellings occur.
- Relate upwellings to predictions of increased plankton productivity and possible increase in available commercial fish.

Procedure:
1. Divide the class into teams.
2. Hand out bathyometric maps or posters and allow time for the students to examine them before giving help in interpretation. (An excellent bathyometric poster of the Gulf of Maine is available from the State Planning Office, State House Station 38, Augusta, ME 04333.)
3. Ask students to locate ocean canyons, walls, peaks, and shallow valleys.
4. Explain what an upwelling is, and ask students to locate places where upwelling might occur. How many places can they find? If they worked for a commercial fishing vessel, where would they go to find the most fish?

Activity 2: The Eerie Light of the Ocean Night

Objective:
Increase awareness of bioluminescence and its possible functions.

Background:
Many people, including the "Ancient Mariner," have observed the phosphorescent seas that occur as a result of light production in the surface waters by billions of dinoflagellates and comb jellies (chicken-egg-size, soft-bodied, slow-swimming animals). *Bioluminescence* is the production of

light by living organisms. The same mechanism is used by some terrestrial organisms such as lightning bugs. A chemical reaction between a substrate called luciferin and the enzyme luciferase produces a molecule that emits light. The color varies from species to species, but tends to be in the blue or green range.

The reasons organisms glow include attracting mates, acting as lures, creating light or camouflage, and confusing predators. However, why or if the glow of some dinoflagellates helps ensure survival is still an unsolved mystery.

Materials:

Some scientific supply houses and toy stores carry glowsticks and chemi-luminescent demonstration kits which contain solutions that produce a blue glow. The glow lasts from 5 to 15 minutes depending upon which kit you purchase. We suggest that you purchase a kit that gives a longer glow. See the Equipment and Material Resource List.

Procedure:

1. Introduce the topic of bioluminescence.
2. Darken the room and begin the demonstration.
3. While the solution is luminescing, discuss "the eerie light of the ocean night." Have any students witnessed the eerie light? What function might it serve for dinoflagellates? (No one knows, yet.) What function does it serve for other marine and terrestrial organisms such as lightning bugs?

Extension Activity:

Write a *Plankton Press* fictional story, poem, or song for children about why the ocean glows at night. You might choose to have them read *Night of the Moonjellies*, by Mark Shasha, a sensitive story of a boy and his grandmother and their adventurous boat trip to see moonjellies (probably bioluminescent comb jellies—like jellyfish).

Question 6
Are There Killer Phytoplankton?

About, about, in reel and rout
The death fires danced at night
The water, like a witch's oils,
Burnt green and blue and white.
  —"The Rime of the Ancient Mariner," by Samuel Taylor Coleridge

All the water changed into blood. The fish died and the river stank.
  —Exodus 7:20–21

OBJECTIVES
Students will:

- Read a short story and discuss the mystery illness of Captain Vancouver's crew.
- Create contributions to *Plankton Press.*
- Analyze a stanza from "The Rime of the Ancient Mariner."

The National Science Education Content Standards addressed in these activities include:

Standard A: Science as Inquiry

Standard C: Life Science—characteristics of organisms; organisms and environments; populations and ecosystems; regulation and behavior

Standard F: Science in Personal and Social Perspectives —natural hazards

Standard G: History and Nature of Science—science as a human endeavor; history of science

Teachers using *Sea Soup: Zooplankton* will also find this unit relevant. The zooplankton mystery is, "Why don't zooplankton that eat toxic phytoplankton pass the toxin to organisms that eat the zooplankton?" The answer is unknown.

Background:

An impressive display in bays, along shorelines, in boat wakes, and in the open ocean, colored waters—red, brown, orange, and luminescent—are collectively known as "red tide." Red tides are one kind of "*bloom*," a rather sudden, huge increase in the number of phytoplankton in a given area. So huge, the water sometimes changes color. Colors come from accessory pigments that mask the presence of green chlorophyll. Most red tides are caused by several types of pigmented dinoflagellates, and most are nontoxic. Many are also bioluminescent.

Blooms may result in illness or death of marine animals and humans. The phytoplankton cells produce a substance *toxic* to some, but not all, marine organisms. Zooplankton that eat phytoplankton do not react to the poison. Fish may die from one kind of bloom but not from another.

The cause of outbreaks of shellfish poisoning in humans remained a mystery until several people died and over one hundred more sickened in California in 1927. A scientist, thinking that the poison was connected to red tides in which the poison shellfish lived and ate, isolated one species of phytoplankton that produced a toxin matching that in the shellfish. Some plankton *are* enemies of humans! In 1965 a species of dinoflagellate was identified as the culprit causing illness in people who ate shellfish. A related dinoflagellate has caused deaths in New York and Nova Scotia.

If a sickness is caused directly by a bacteria, then cooking will kill the bacteria. However, if it is caused by a chemical poison inside the organisms the mussels or fish eat, then it is unlikely that cooking will destroy the poison. Some people like taking the risk. Most of us do not! The only guarantee that shellfish are safe to eat when there has been a bloom is a test done by a Health Department.

Red tides occur anytime from spring through fall, but are most common during late summer. Blooms often occur when nutrient levels are high and physical forces such as wind or heavy rains occur. Wind may blow the bloom into windrows or patches. Humans may exacerbate the normal process of red tides. Red tides occurred historically (as mentioned in Exodus of the Old Testament), but the frequency seems to be increasing. Recent evidence indicates that some blooms are perhaps triggered or intensified by high nutrient levels in runoff. Salinity and temperature also play key roles in blooms.

Research indicates that excessive nutrient (fertilizer) runoff into coastal waters has significantly affected many coastal ecosystems. Loss of dissolved oxygen in waters such as Long Island Sound, Chesapeake Bay, and the Gulf of Mexico may accelerate the frequency of harmful blooms. Biotoxins produced by marine phytoplankton can even affect fishery health. In the early

1990s, the scallop fishery in Peconic Bay, New York, was devastated by a brown-tide bloom. Blooms cannot be predicted reliably. New methods for detecting these toxic blooms must be developed as well as accurate prediction capabilities.

Diving Deeper: **Paralytic Shellfish Poisoning**

*P*aralytic shellfish poisoning (PSP) occurs on coasts throughout the world. PSP is primarily passed to humans through shellfish. Other organisms that feed on the same toxic phytoplankton do not cause the same reaction in humans when eaten. In the worst cases of PSP, respiratory paralysis and death occur within 2 to 12 hours. One of the toxins, saxitoxin, has been isolated. It is a nerve poison that blocks transmission of impulses from nerve to muscle.

Research on the causes of PSP indicate that dinoflagellate Gonyaulax blooms occur during long spells of warm, clear, sunny weather following heavy rainfall and high fresh-water runoff.

Diving Deeper: **Shellfish and Phytoplankton**

*C*lams, mussels, and oysters feed on zooplankton, phytoplankton, and other very small organisms. They obtain their food by pumping water into their shells and over their gills where food is trapped and directed towards the mouth. These three shellfish live in the intertidal zone near the water's surface where phytoplankton abound. Scientific studies show that mussels can become poisonous within a few days after a dinoflagellate outbreak begins. Soon after a bloom, mussels rid themselves of the poison. However, other kinds of shellfish keep the poison for as long as two years after the initial outbreak. Strangely, the poison does not harm the shellfish. It is poisonous to some fish, such as herring, and mammals (cats are particularly sensitive).

Shellfish and fish may also be killed indirectly by phytoplankton. After a bloom, the number of dinoflagellates crashes. The dead cells decay. The process of decomposition of dinoflagellates by bacteria depletes the water of oxygen. Organisms such as fish may then die due to lack of oxygen. It is a vicious cycle!

Diving Deeper: **Poisonous Diatoms**

*S*ome species of diatom cause amnesiac shellfish poisoning (ASP) in humans. Pseudonitzschia produces domoic acid causing ASP. Other diatom species cause fish kills by producing mucus that clogs fish gills or by irritating fish gills.

Activity 1: **The Mystery of Captain Vancouver's Sick Crew**

Objectives:
Students will listen to a short story about the illness and death of some coastal explorers, and analyze and discuss "what killed the crewman."

Materials:
- One copy of the story and questions for each group.

Procedure:

1. Make enough copies of the story and questions for each group.

2. Divide the class into groups of three or four students.

3. Read the short mystery story to the class or distribute one copy per group for one student to read to their group.

4. Allow 15–20 minutes for students to discuss the questions and try to solve the mystery about what killed the fish and sickened the people.

5. Refocus the class for a culminating discussion of the questions with members of each group contributing to the discussion.

6. Assess their group's insight and understanding of the questions and answers.

7. Now, assign the section in *Sea Soup: Phytoplankton,* "Are there killer phytoplankton?"

The Mystery of Captain Vancouver and His Crew

In June 1793 Captain George Vancouver and his crew sailed the central coast of British Columbia. They anchored in a familiar bay, and then rowed small boats to shore to explore the area again.

Four sailors ventured into a small rocky cove. It was breakfast time and they were hungry. Multitudes of mussels grew attached to the intertidal rocks by strong threads. The men loved mussels for breakfast, or any other meal! They'd eaten some mussels from the same coastal area several days in a row.

So, the men pried the mussels off the rocks, opened the shells, and slurped the mussels down their gullets, live and whole.

Within a few minutes, the men's lips and fingertips were numb.

Then, their arms and legs became paralyzed.

Dizziness and nausea set in.

One man died.

The others began to run, jump, and exercise!

They lived.

Discussion Questions:

1. What was happening?

2. What killed the sailor? Something in the water? Something they ate on board the ship? Too much exercise rowing through the surf?

3. Why did the other three sailors live?

4. Since they had eaten raw mussels for breakfast on other days, why did they get sick on that day?

5. What event happened to change the mussels?

6. If you were one of the explorers, what would you have named the cove?

7. If you were offered raw oysters, mussels, or clams, would you eat them?

8. Do you think cooking the shellfish might have made a difference in the end of the story? Why or why not?

Activity 2: **Be Creative with Plankton!**

Objectives:

Students will explore phytoplankton through imagery and writing, drawing, etc.

Procedure:

1. Read one or both of the quotes printed at the beginning of this section to the class. Ask them to visualize the image created by the words.

2. So far your students have explored and learned about many aspects of phytoplankton. Now is a good time to give an assignment in which they expand this topic in another direction and use their own particular skills to demonstrate this knowledge.

• Ask your class to work independently in creating a contribution to the *Plankton Press.* A poster, poem, story, cartoon, news flash ("Upwelling Is Coming!"), health update, cinquain, diamante, or a song relating to plankton, ocean food webs, blooms, etc.

Extension Activity (most appropriate for grades five and six):

Locate and read the poem, "The Rime of the Ancient Mariner," by Samuel Taylor Coleridge. Ask the students to take notes about any biological phenomena referred to in the poem. Then, hand out copies of the Coleridge quote from the beginning of this section about blooms. Discuss the story and the setting of the poem. Discuss the context for the stanza quoted. How does that stanza relate to the topic of phytoplankton?

Question 7
Have You Thanked a Phytoplankter Today?

Phytoplankton play important roles in our lives. They produce 95 percent of the oxygen available for use by ocean organisms and contribute over 50 percent of the earth's oxygen. Without phytoplankton, life would not exist as we know it. Without oxygen generated by phytoplankton, ocean organisms—including zooplankton—would die. The whole ocean food chain would collapse. Terrestrial organisms would suffer greatly. Phytoplankton impact the earth's environment in other ways, too. Our civilization depends on petroleum-based energy sources such as oil and gas. They formed from phytoplankton and plant material eons ago.

OBJECTIVES
Students will:
- Explore the roles phytoplankton play in our lives.
- Explore the role we play in the lives of phytoplankton by modifying the atmosphere (greenhouse effect).

The National Science Education Content Standards addressed in these activities include:
Standard A: Science as Inquiry
Standard B: Physical Science—properties of matter
Standard C: Life Science—Organisms and environments
Standard F: Science in Personal and Social Perspectives—populations, resources and environments; science and technology in society

If you are just using Sea Soup: Zooplankton with your class, please do not skip over this section because it seems too phytoplankton oriented. You will find that most of the activities in this section are relevant. Remember to emphasize that zooplankton are a vital link in the ocean food chain because they eat phytoplankton and are critical food for other animals. Below is the pertinent text from *Sea Soup: Phytoplankton.* You have our permission to copy it for use in Activity 1.

Phytoplankton protect us from ourselves. The earth has been getting warmer over the last 100 years, and many people blame it on the "greenhouse effect." They believe that the burning of fossil fuels and rain forests releases carbon dioxide into the air which traps heat in the upper atmosphere, causing global warming. Each year, phytoplankton take nearly half the carbon dioxide we release by burning coal, oil, and gas and turn it into shells for phytoplankton like coccolithophores. When they die, the carbon sinks to the bottom of the sea inside them. Phytoplankton use 3 billion tons of carbon dioxide each year, slowing down the warming of the earth. Phytoplankton also help replace the ozone layer. (Chemicals we use in hair sprays and refrigerators have helped make the ozone layer thinner.) The ozone layer helps protect us from the harmful UV (Ultraviolet) rays of the sun that can cause sunburn and even skin cancer.

Many of us heat our homes with the remains of ancient phytoplankton. When they died, the oil-filled phytoplankton sank to the ocean floor and were eventually buried under layers of mud, sometimes thousands of feet thick. They changed into oil deposits over millions of years. Eventually, geologists drilled into the ocean floor and brought the phytoplankton, changed into oil or natural gas, back to the surface of the sea. Most oil deposits are found under the ocean or under land that once was covered by the sea.

Sometimes as tankers move the oil across the ocean, the ships run aground on rocks or reefs and spill their contents into the ocean. Oil coats the seashore, the ocean surface, birds, otters, and other marine life. It poisons their habitat. Oil-eating bacteria are brought in to clean up the oil, but they leave behind carbon dioxide. Phytoplankton are brought in to take away the carbon dioxide.

In recent years people have added to what the sea can provide through *aquaculture*, raising fish, shellfish, and other foods from the sea in shallow bays. Phytoplankton are used to feed animals being raised in aquaculture operations, or farming the sea. Products made from phytoplankton also filter swimming pools, distill fruit juice, wine, and beer, put the polish in toothpaste, and keep dynamite from exploding too soon. Perhaps most important, phytoplankton help us to breathe. About half of the world's oxygen comes from phytoplankton. That means every other breath you take is thanks to phytoplankton.

Activity 1: Explore the Interaction Between Humans and Phytoplankton

Objective:
Students will:
• Dive into specific examples of how phytoplankton are involved in our lives, and how we are involved in the lives of phytoplankton.
• Design and create Plankton Posters on the theme, "Have you thanked a phytoplankter today?"

Background:

Sea Soup: Phytoplankton introduces many examples in which phytoplankton play an important role in our lives, and examples in which humans' actions create potential problems for phytoplankton. Depending on the grade level of your class, you may choose to discuss each topic based on the book, or you may choose to expand the topic through web searches and other readings.

Topics can include, but not be limited to, the relationship between phytoplankton and one of the following:

Ozone layer
Global warming
Greenhouse effect
Coccolithophore shells, carbon dioxide, and oil
Oil/gas
Oil spills
Oxygen
Diatomaceous earth

Materials:

Paper for posters
Markers

Procedure:

1. Assign the pages in *Sea Soup: Phytoplankton* relating to "Have you thanked a phytoplankter today?" Ask the class to make a list of examples mentioned in the reading in which humans "use" phytoplankton and a list of examples in which humans' actions may be causing changes that may impact the health of phytoplankton.

2. Discuss the reading.

3. Ask the students to work independently to design a poster using the theme, "Have you thanked a phytoplankter today?" They can emphasize the positive uses or the negative human actions. This activity could be undertaken jointly with the art teacher and expanded to become a banner.

4. If wall space is available outside the classroom, post the posters so other school children can see them.

Extension Activity:

Students choose one of the topics for an investigative report. The resulting report should be written in an engaging way as a good newspaper reporter would write. The reports will be shared through the production of the *Plankton Press.*

Activity 2: The Fisherman's Song

Objective:

Students analyze the anonymous poem, "The Fisherman's Song" and discuss the questions in writing or in class.

The Fisherman's Song

When haddocks leave the Firth of Forth,
And mussels leave the shore,
When oysters climb up Berwick Law,
We'll go to sea no more,
No more,
We'll go to sea no more.

O blithely shines the bonnie sun
Upon the Isle of May,
And blithely rolls the morning tide
Into St. Andrew's bay.

Discussion:

1. What clues tell you in which country this poem was written?
2. What is a "firth?"
3. Look on a map or in an atlas to locate the Firth of Forth, Isle of May, and St. Andrew's Bay. What is the name of the large body of water nearby?
4. Can anyone find out what Berwick Law is? (Possible answer: Berwick is a county in southeastern Scotland, so perhaps the poem refers to a local law about oystering. However, a remote meaning of "law" is an allowance of time, like a head start, given a quarry or a competitor.)
5. Why does the fisherman say, "We'll go to sea no more?"
6. What factors might cause the haddocks, mussels, and oysters to "leave the shore?" What human actions might have caused the loss of the animals? What role might phytoplankton and zooplankton have played in the loss of the animals?
7. What does the second verse imply? (Answer: No one seems aware that there are problems—maybe loss of plankton, over-fishing, pollution, or red tide—causing the loss of the fisheries.)

Activity 3: Carbon Dioxide, the Ocean Sink, and the Greenhouse Effect

Objective:

Students create miniature greenhouses and measure the effect of light on air temperature in the greenhouses. (This activity is more appropriate for upper elementary and middle school.)

Background:

Since the beginning of the industrial revolution in the late eighteenth century, the amount of carbon dioxide in the atmosphere has increased by about 30 percent. This increase may be affecting our climate. Carbon dioxide, water vapor and some other gases absorb infrared radiation from the surface of the earth and trap heat within the atmosphere. They are commonly called "greenhouse gases" because they trap heat like a greenhouse traps heat.

As the human population increases, burning of fossil fuel increases, and the amount of atmospheric carbon dioxide also increases. Increased greenhouse gases may result in climate changes. Temperatures may increase on a scale of centuries rather than over millennia, as they did in the past.

Oceans store huge amounts of heat as well as dissolved carbon dioxide from the air. This is called an "ocean sink." The oceans act as buffers against climate change. However, increased solar heating and evaporation produce surface waters that are warmer and saltier than normal. Eventually this influences global ocean circulation in the deep, slow-moving currents. The eventual impacts of ocean changes are unknown.

Scientists study records of past oceanographic and climatic conditions in seasonal variability of sea surface temperature, salinity, nutrient content, and other elements by examining natural records in deep-sea sediments, ice, and corals. Questions to be answered by scientists include:

- How have these factors varied in the distant past?
- How has this variability changed over time?
- What is the response of the ocean to atmospheric changes?
- What are the issues about which we should be concerned?
- What can we do to reduce human impact on the atmosphere and oceans?

Materials (per pair of students):

Copy of Student Discovery Master 7: The Greenhouse Effect for each student

2-gallon jars or 2-liter clear soda bottles with labels removed and slanted portion of top cut off

1 piece clear plastic wrap about 12" x 12"

2 thermometers

1 rubber band

2 cups soil or sand

Desk lamp with at least a 100-watt bulb

Tape

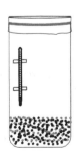  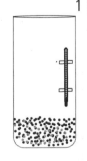

Procedure:

1. Collect and organize the materials for this investigation including data sheets for each group.

2. Divide the class into pairs.

3. Give a simple introduction to the investigation. State that the students will construct simple gallon-jar greenhouses and record the temperature in each jar for fifteen minutes. After the class has collected the temperature data, introduce the topic of "the greenhouse effect" by discussing how carbon dioxide in the atmosphere acts as a heat trap for the sun's energy reflected from the earth just as the plastic over the jar or glass in a greenhouse traps the heat energy.

4. Hand out the simple instructions and the data sheet.

5. After the students complete collecting data, allow enough time for them to graph the data and answer the questions on the student handout.

6. Discuss the results and use the results as an introduction to "the greenhouse effect."

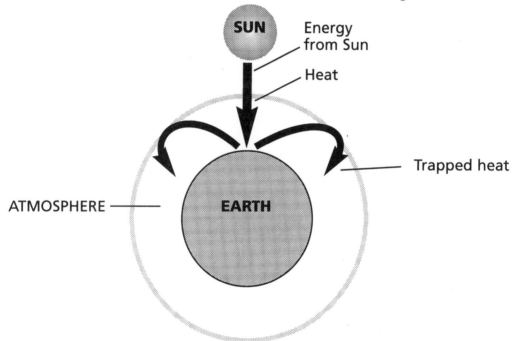

The Greenhouse Effect

Extension Activity:

Research and/or discuss the possible effects of rapid and significant global warming. Among the possibilities are the impacts on health (U.S. might be at risk for tropical mosquito-carried diseases); agriculture (the growing season in northern areas would be lengthened); forests (forests pests would thrive); range lands (productivity would decline from drought and fires); wildlife (many species would become extinct because their habitats would shrink, shift, or disappear); and oceans (currents and fishery areas could change and coastal areas could flood).

Student Discovery Master 7:
The Greenhouse Effect

Name _____ **Date** _____

Class _____ **Names of partners** _____

Challenge:

To model the greenhouse effect and record temperature data when chambers are exposed to light.

Introduction:

The earth is surrounded by a layer of air that acts a lot like the glass in a greenhouse. The layer of air allows the sun's light and heat to come in but does not allow it all to get out. This is important because without the layer of air trapping some heat the earth would be much colder. The trapped heat has raised the temperature of the earth about 60° F higher than it would be otherwise. The naturally occurring atmosphere has created a temperature balance, and living things have adjusted to it.

Unfortunately, because the amount of one of the "greenhouse gases," carbon dioxide, has increased about 30 percent in the last 100 years or so, the atmosphere may be trapping too much heat. Some scientists are afraid that the earth may some day be too hot.

Why is the amount of carbon dioxide in the atmosphere increasing? As humans burn fossil fuel such as oil, coal, and gas, and burn forests, large amounts of carbon dioxide are released. The oceans store huge amounts of dissolved carbon dioxide from the air, and they store heat. But the oceans have not been able to store the extra amounts released by burning so much fossil fuel and forests.

The greenhouse effect and global warming are new areas of scientific study. We do not have all the answers yet.

Instructions for Assembling Greenhouses:

1. Check to make sure you have all the materials listed.
   - 2-gallon jars or 2-liter clear soda bottles with labels removed and slanted portion of top cut off
   - 1 piece clear plastic wrap about 12" x 12"
   - 2 thermometers
   - 1 large rubber band
   - 2 cups soil or sand
   - Desk lamp with at least a 100-watt bulb
   - Tape
2. Check the thermometers to be sure they read the same temperature value.

If they do not, make a note to add or subtract the error from the temperature readings made with that thermometer.

3. Place 1 cup of soil in the bottom of each jar.

4. Tape a thermometer on the inside of each jar with the scale facing out so you can read it.

5. Cover one jar with plastic wrap and secure with the rubber band.

6. Set the jars an equal distance from the light so it shines into the sides of the jar. DO NOT TURN THE LIGHT ON YET. Make sure the thermometers face away from the light.

7. Which jar is the control?

8. Set up a table for recording your data. One of you will read the temperature and the other will record it.

9. Make a prediction about which bottle will become the warmest.

10. Will the temperature level off as long as the light shines?

11. Now, turn on the light and record the temperatures every 2 minutes for 15 minutes on the data graph.

12. After 14 minutes, stop taking readings. Make graphs of your data. Be sure to fill in blanks and scale for temperature.

Answer the following questions and hand in to the teacher:

1. What trends do you observe in each bottle?

2. Was there a difference between the temperatures in each jar? Describe.

3. Did the temperature in either jar level off? Why or why not? (Hint: Does any heat escape through the glass?)

4. How does this experiment correspond to the real earth?

Activity 4: Investigating the Properties of Diatomaceous Earth

Objectives:

Students will discover answers to the questions:

• What is diatomaceous earth?

• What does it look like?

• How can it be used?

Materials:

Diatomaceous earth

Schoolyard soil

Spoon

Funnel

Filter paper

Cheesecloth

Fruit juice with pulp

Cloudy water

Pond water

Procedure:

1. Make copies of Student Discovery Masters 8 and 9.

2. Assemble materials listed above.

3. Divide class into pairs.

4. Hand out Student Discovery Master 8 and ask students to follow the directions.

5. Discuss the results.

6. Students then proceed with Student Discovery Master 9.

Student Discovery Master 8:
Exploring Diatomaceous Earth

Name _____ **Date** _____

Class _____

Introduction:

You have learned a lot about different kinds of phytoplankton and the roles they play in marine food webs and ecosystems. Phytoplankton also play many important roles in our daily lives.

Diatoms—the flattened, round, pillbox-shaped phytoplankters—use carbon dioxide and calcium to make their hard, sculptured, two-part shell. When diatoms die, the soft, watery cell decays or is eaten by other organisms, but the shell sinks to the bottom of the ocean. Billions and billions of these shells accumulate and eventually form what is called "diatomaceous earth." Over millions of years, as the earth's surface changes and the sea level rises and falls, large pockets of diatomaceous earth are exposed. Humans have used diatomaceous earth as insulation on the roofs of houses in the hot desert, and as filters at breweries and fruit juice companies, and in swimming pools.

Challenge:

Investigate some properties of diatomaceous earth.

Materials:

Sample of diatomaceous earth
Sample of schoolyard soil
Spoon
Funnel
Filter paper
Cheese cloth
Cloudy, natural fruit juice with pulp

Procedure:

1. Put a few spoonfuls of diatomaceous earth into a shallow container.
2. Put a few spoonfuls of schoolyard soil into a shallow container.
3. Feel the two earth samples with the tips of your fingers.
4. Answer these questions:
• Describe the visual appearance of diatomaceous earth. How does it compare with the schoolyard soil?

100

Student Discovery Master 9:
Design an Experiment

Name _____ **Date** _____

Class _____

• Describe the difference in the feel of the two earth samples. Which one is coarser?

You and your partner own a fruit juice company called Jolly Juice. You pride yourself on producing tasty juices with natural sweetness and flavor. However, customers have e-mailed messages to you saying, "Yes, I used to like juice with pulp, but now I want clear, colorful, tasty juice." How can you redesign your production line to produce clear juice without any suspended particles?

Challenge:

Using some of the materials at your station, design and carry out a test for determining which materials and method would work the best to solve your problem.

Procedure:

1. State your specific question or hypothesis being tested.

2. Describe your procedure here step by step. Be sure to consider questions such as, "How will you measure which filter is best?"

3. Sketch your setup here.

4. Record your observations.

5. Record your results:
• Which setup (filter) resulted in the product that meets your requirements?

• Why did that setup work the best?

• If you were to repeat this investigation, what would you do differently?

6. Now, your teacher may request that you research the uses of diatomaceous earth through the Internet or library.

Extension Activity (requires compound microscope; may be set up as a demonstration):
How does diatomaceous earth differ from schoolyard soil?

Procedure:
1. Set up a compound microscope.
2. Prepare a slide by putting a small speck of diatomaceous earth in the center of the slide; add one drop of water on the speck and slowly place one coverslip on the drop.
3. Be sure the low-power objective lens is in place. Now, put the slide under the lens and center the speck directly under the objective lens.
4. Carefully focus the image. Remember, the diatom shells are very small and may look like dust particles. If you are seeing a big, brown blob, then you have too many shells on the slide. Wash and dry the slide and try again using a very small amount of the diatomaceous earth. Ask your teacher for help, if necessary.
5. Both you and your lab partner should observe the slide.
6. Discuss what you see. Answer the questions. Now, carefully switch to the high-power lens, and refocus using the fine adjustment knob only. Again, ask for help.
7. Discuss what you see with your partner. Answer the questions 1–4.
Now, prepare a slide of schoolyard soil by following steps 2–6. Answer Question 5.
8. When you finish your observations, wash the microscope slide and dry it. Switch the objective lens to low power and put the microscope away.

Discussion Questions:
1. Describe the shells. Sketch a few representative shells.

2. Do all the shells look alike? How do they differ?

3. Do the shells look like the drawings and photos you have seen in previous classes? Why or why not?

4. Describe the schoolyard soil as it appears under the microscope. Compare it with the diatomaceous earth.

How Do We Know So Much About Things Most of Us Have Never Seen?

Advanced scientific research on plankton, especially the microscopic world of phytoplankton and zooplankton, requires meticulous fieldwork and sophisticated laboratory equipment. Technological advances such as satellite arrays, flow cytometers, DNA analysis, and sophisticated fluorescence microscopes allow scientists to answer questions. However, much can be learned, especially by students using simple, handmade equipment and their own senses.

OBJECTIVES

Students will:
- Build field equipment
- Go on a field trip
- Examine and analyze their samples
- Meet a marine or freshwater scientist

The National Science Education Content Standards addressed in these activities include:

Standard A: Science as Inquiry

Standard C: Life Science—organisms and environment; ecosystems

Standard E: Science and Technology—abilities of technological design

Standard F: Science in Personal and Social Perspective—science and technology in local challenges

Standard G: History and Nature of Science

Assign the class to read the *Sea Soup: Phytoplankton* section about "How do we know so much about things most of us have never seen?" After they complete the reading, they may be anxious to do some fieldwork and investigations.

This activity is appropriate for investigating zooplankton as well as phytoplankton. *The text in* Sea Soup: Phytoplankton *discusses some equipment used to investigate plankton.*

Activity 1: Create Sampling Equipment and Water Quality Monitoring Equipment

Characteristics of a water sample can be measured using a variety of instruments, some of which are technologically complex and others which can easily be made in the classroom or purchased. The instruments used in this activity include buckets, Secchi disks, a plankton net, a hydrometer, and a thermometer.

• Buckets
Ask students, cafeteria workers, or a building supply store for white 5–gallon buckets. They are handy as equipment carriers and for obtaining water samples from bridges or docks. The lids can be used for the Secchi disks.

If you choose to use a bucket to obtain water samples, attach a long rope to the handle so the bucket can be lowered down into the water or thrown out from the water's edge.

• Secchi Disks
On your field trip you will use a Secchi disk to measure turbidity, the relative clarity of water.

Turbid water appears murky. Solid particles such as eroded soil suspended in the water can block out light that phytoplankton and aquatic plants need. Suspended materials can absorb heat from sunlight raising the water temperature. As water becomes warmer, its ability to hold oxygen decreases. When the dissolved oxygen level drops, many animals cannot survive and the whole system is impacted. Water rich in plankton may also appear turbid because the plankton absorb light.

Materials:
Each group making a Secchi disk will need:
White or opaque plastic lid about 6 to 8 inches in diameter
String (about 15 feet or more if the water you will sample is very clear)
Yarn
Yard stick
Eyebolt with two washers and two nuts
Hammer
Large nail
Piece of junk wood
Black waterproof marker

Procedure:

A. Make a Secchi Disk.

1. Use the hammer and nail with the junk wood as backing to punch a hole in the center of the plastic lid.

2. Color the lid (disk) as shown in the illustration.

Be sure to color the *underside* of the lid.

3. Put a nut and then a washer on the eyebolt.

4. Insert the eyebolt into the hole in the disk.

Be sure the colored *underside* of the lid faces up.

5. Add the other washer and nut in that order to the eyebolt and tighten so the disk is secure on the eyebolt.

6. Tie one end of the string to the eyebolt.

7. Measure the string out 10 inches from the eyebolt and tie a piece of yarn tightly at that point.

Continue measuring and tying a piece of yarn every 10 inches for the length of the string.

B. Now the Secchi disk is ready to use on the field trip.

1. To measure the turbidity of the water, the disk is slowly lowered into the water until you cannot see it.

2. Then raise it slowly until it just appears.

3. Find the point at which it just disappears.

Note which piece of yarn is just above the water's surface.

As you pull the disk out of the water, count the number of yarn pieces between the one just above the water's surface and the disk.

That number represents the turbidity reading.

4. Repeat steps 1 through 3 three times at the site.

5. Move to other locations and determine the turbidity.

Again, repeat three times at each site.

6. Be sure to keep a record of your results and the location of the sampling sites.

7. If your Secchi disk is still visible when it reaches the bottom, record the number as all the yarn on the string between the bottom and the water's surface and add a plus sign after the reading.

8. Average the number for each site by adding the turbidity readings and then dividing by the number of times you made a measurement at a particular site. This is the data point you will enter for each site on the class "Field Trip Observations" chart back in your classroom.

• PLANKTON NET

To sample the water for plankton and suspended solids, students will use plankton nets made in the classroom.

Materials:

Each team making a net needs:

1 Coat hanger or 3–foot piece of heavy wire

1 Nylon stocking or other fine mesh material

Needle and thread

Scissors

Pliers

Wire cutter

Wax pencil

A clean, clear jar

Clipboard or notebook

Pencil or permanent marker

Procedure:

1. Unwind the coat hanger using the pliers. Straighten out the wire.

2. Bend about 10 inches of the wire to form a circle leaving the rest of the wire to form a handle.

3. Bend the remaining wire back on itself to form a handle.

4. Fold the top of the nylon stocking over the circle and sew around the rim.

5. Tie a tight knot in the stocking about a foot from the rim. Cut the extra stocking off. The net is ready to use.

You can obtain the sample in several ways—either tow the nets behind a boat, toss the net out into the water and pull it into the dock or shore, or slowly pour many buckets of water through the net. When you finish sampling the water, gently wash the filtered material to the bottom of the net by pouring a little water down the sides of the net. Then reverse the net and wash the sample into a clean jar and tighten the lid for transport back to the classroom. Take a quick look at the sample to see what is visible. Make notes on the clipboard. Label the jar.

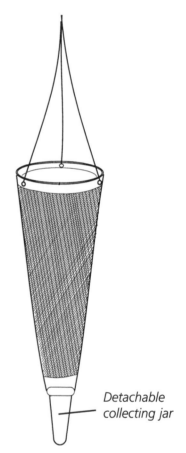

Detachable collecting jar

Note: You may choose to purchase a fine-mesh plankton net from a biological supply company. It is easier to tow behind a boat and the finer mesh will capture more plankton. However, the "stocking" net works well.

- HYDROMETER

  If your field trip is to freshwater sites, then skip making this equipment. If you will be sampling marine sites, use store-bought hydrometers or make simple hydrometers by putting a lump of glue on the end of a straw to seal it and to give it some weight.

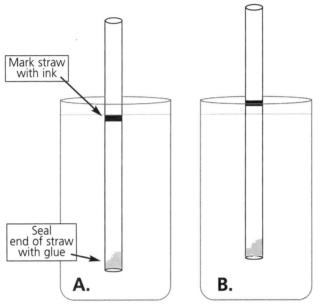

Mark straw with ink

Seal end of straw with glue

A. *Straw hydrometer in water*

B. *Straw hydrometer in saline*

Put the straw hydrometer into a sample of water of known density (use your store-bought hydrometer to test). Make a line with permanent marker

Store-bought hydrometer

at the point on the straw where it floats in the water. Label the spot with the density. Then, put the straw in another sample of water of known density and label. Repeat with freshwater or ocean sample. Now you have a straw hydrometer that will give you relative readings. Remember, relative readings in which you compare freshwater with marine sites can be just as meaningful to students as obtaining scientifically accurate readings using expensive, delicate equipment.

- THERMOMETER

  Water temperature can be an important factor in water quality. All other factors being equal in water samples from several places, a relatively higher temperature may indicate lower dissolved oxygen than in cooler waters. Lower oxygen may result in death for animals, especially fish.

  Purchase several inexpensive thermometers from an aquarium store or biological supply company. Check all thermometers for relative accuracy. If some thermometers read a degree or so different from the majority, make a note on the thermometer with permanent marker indicating how many degrees need to be added or subtracted from that thermometer's reading in order to make it comparable with the other thermometers.

Activity 2: Can You Find Phytoplankton in Your Own Backyard? A Field Trip

Whether you walk to a small creek flowing along the school property or take a chartered "Day on the Bay" boat trip, students can experience the excitement of scientific research and water quality monitoring. Choose a safe site for water sampling. If wading into the water or dropping a bucket over the side of a bridge, boat or dock is not possible, you can toss a sampling bucket (with rope attached!) out from the shore.

Materials for field trip:
>   For each team:
>   Secchi disk
>   Thermometer
>   Hydrometer
>   Bucket with rope attached to handle
>   Plankton net
>   Plastic jars with lids for plankton samples
>   Magic marker or wax pencil
>   Instructions/field data sheet on clipboard
>   Clear plastic bag
>   Pencil (pen ink runs if it gets wet; field scientists use pencils or permanent ink)
>   Large garbage bags for litter collection
>   First aid kit
>   Old sneakers or reef runners or hip boots
>   A few old towels

Procedure:
>   1. You may choose to develop a field trip data log for use by each team. We do not include one in this activity guide because field trip circumstances will vary from school to school.
>   2. Divide your class into teams. Depending on what you include in the field trip, each team might do the same set of activities which will allow comparison of several teams' data. Or, each team might be assigned one particular task and then share the results with the whole class.
>   3. Make assignments for each team. All teams should have an appointed leader who oversees the activities for his/her team. Another student should carry a clipboard and record results. (Put the clipboard in a clear plastic bag large enough so that the student can use a pencil to write on the board inside the bag in case of rain.)
>   4. Go on the field trip and have fun!

Note: If time does not allow examination of the samples and discussion of the data on the same day of the field trip, be sure to keep the water/plankton samples at approximately the same ambient temperature at which they were obtained. **IMPORTANT:** *Remove or loosen the jar tops and if possible keep low-level light on the sample.*

Activity 3: **What's in Our Water Samples?**

Back in the classroom, pool the data:

1. Enter all the numerical data from all teams on the blackboard. See suggested format. The Field Trip Observations chart should include:

• List of the sampling sites. For example, "Site C: End of dinghy dock at Chick's Marina."

• Team data from each site.

2. Discuss any major differences within all the teams' data for each factor sampled. For example, Site B Temperature sampled by all teams at the same site but one team got higher readings. If significant differences exist, discuss possible explanations for the differences, e.g., sampling error, faulty equipment. If any mistakes were made by the team, you will have to decide whether or not to eliminate that data. (Note: For upper elementary classes, this is an opportunity for the teacher to introduce the concept of sampling error and to discuss how you know if a difference among data is real or due to an error, and how you decide to whether or not to eliminate data.)

Field Trip Observations

| Team No. | Site A | | | Site B | | | Site C | | |
	Temperature	Salinity	Turbidity	Temperature	Salinity	Turbidity	Temperature	Salinity	Turbidity
1									
2									
3									
4									
Average for each site									

Examine water collected on the field trip. Also examine samples collected by the teacher from other sources or obtained from a biological supply house. *(Note: Since samples and magnifying equipment will vary in different schools, we have not produced this activity as a reproducible handout for your students. You may choose to do so.)*

Materials:
  Slides
  Eyedroppers
  Magnifying glasses
  Microscope
  Petri dishes
  Plankton net samples

Procedure:
  1. Look at the sample jars and note any visible movement, etc.
  2. Using an eyedropper, sample the material on the bottom of the jar.
  3. Put one drop on a slide and gently place a coverslip on the drop.
  4. Examine under a microscope if available.
  5. If other magnifiers are available, place a sample in the appropriate containers and observe with the magnifier. Take notes on your observations and sketch what you see. Ask for help from teammates and your teacher.
  6. Use available guides to help identify the organisms and debris in your samples.
  7. Repeat by taking samples from elsewhere in the samples and by catching visible zooplankters.

Instruct the students to include their plankton observations in their Field Trip Report.

Homework Assignment: Field Trip Report
Ask the students to:
1. State the purpose of the field trip.
2. List the factors you measured.
3. Make a chart with all class data.
4. Highlight your team's data in appropriate columns.
5. Compare data for each factor from each site.
Make a bar graph for data if you think it helps illustrate the results.
6. Are there differences between the sites?
Discuss the differences and why they exist.
7. What conclusions can you draw from the data?
8. What other factors do you think should be included in monitoring water quality?

Activity 4: Scientist in the Classroom

Arrange for a classroom visit from a local water quality person or a marine or freshwater scientist who studies water quality, plankton, fisheries, oceanography, or a related field. Emphasis should be on what that person does and the tools necessary for doing the work.

Extension Activity: Town Meeting/Role Playing

In some coastal areas of the country, marine aquaculture has become a significant contributor of seafood to the marketplace. Some issues relating to aquaculture include the impact on local ecosystems and the impact on the local economy. Students can research the topic on the Internet and in publications from federal and state agencies. Then, you may choose to divide the class into teams such as aquaculturists, investors, marine scientists, environmentalists, developers, tourist industry people, and local citizens. One student acts as the moderator of a town meeting or a planning board meeting at which the proposal for an aquaculture facility is presented. All parties present their comments and arguments for or against the proposal. Perhaps some will present possible compromises.

Culminating Activities

• Review the contents of the Sea Soup kettle (on the bulletin board) and discuss the recipe!

• Review the results of the phytoplankton and fertilizer experiment.

• To help assess the value of the Sea Soup unit for your students, ask them to write an essay about the meaning of the following quote:

"There, said he, pointing to the sea, is a green pasture where our children's grandchildren will go for bread."

—Obed Macy, *A History of Nantucket*

Enrichment Activities

You have investigated and learned a lot about plankton. Now develop a new superspecies of plankton. Describe how it is different in form and function from some of the plankton you have "met" in the Sea Soup unit.

• You are a reporter covering the top ocean story of the year. What is it? (Be creative!)

• Design a T-shirt with a message about our oceans.

• What is the most important thing to you that you would like to tell people about the ocean?

• If you could change one thing that humans do which impacts the ocean, what would it be?

• If you could take a tiny submarine anywhere on earth, where would you go? Why?

• Design a poster, cartoon, or bumper sticker that makes a statement about one of the following:

       Importance of plankton

       Red tides

       Ocean food web

       Plankton, the food of the future

       An issue relating to human impact on natural resources

       An issue relating to human impact on the environment

• Discuss what actions students could take to raise awareness of:

       The importance of phytoplankton for the welfare of all other living things

       The importance of healthy oceans

       The role oceans play in moderating the earth's temperature and carbon dioxide levels

       The impact of humans on water quality

       The impact of water quality on plankton and the food web, and eventually humans and the health of the whole earth

• *Discuss the following questions with advanced students. Included is some helpful background information.*

1. How do oceans contribute to the earth's biodiversity and natural resources?

Biodiversity is the collection of all organisms and the nonliving components of their ecosystems occurring in a geographically defined area. The loss of one species could precipitate the loss of many more that depend upon it.

Because of the difficulties in sampling the oceans, large areas of the seas have never been explored. Routinely new species are discovered. Some marine biologists suggest that in the deep sea alone, there may be as many as ten million unknown species. Human actions may already be destroying multitudes of species yet undiscovered. Perhaps some would be potential medicines, like so many plants of the tropical rainforests. Recently a source of a new anti-inflammatory medicine for the treatment of arthritis was discovered in the Caribbean sea whip, a soft coral.

Harvesting of commercially important species depends on the ability to accurately predict future abundance and distribution. Without information on population size, reproductive success, number of surviving larvae, and how that relates to oceanographic conditions such as temperature, water clarity, salinity, light, current speed and direction, bottom composition, and food, human harvesting of the sea may come to a stop.

Mining possibilities in the deep seas are just now being explored. Discovery of deep hydrothermal vents and the recently developed technology that allows mineral rich chimneys to be brought to the ocean's surface have excited potential exploiters of the minerals of the deep seas. At present, no regulations exist to control mining of the earth's deep seas.

2. How are humans causing potentially irreversible changes in the biological diversity of marine organisms?

Pollution, commercial fishing, physical alterations to the coasts, introductions of non-native species, and activities contributing to global climate change contribute to irreversible changes in the biological diversity of marine organisms.

Serious decline in the abundance of most commercial fisheries, loss of species with potential for use in biotechnology (development of medicines), reduced aesthetic and recreational value of coastal areas, changes to the structure and function of marine ecosystems, and potentially harmful effects on human health result from human activities on our oceans and coastlines.

3. Why are the oceans vitally important to every living organism on earth?

Oceans influence climate and weather, are an important source of food for the world's population, contain information about the nature and history of the earth, and are the final resting place of many pollutants. We must continue intensive studies of the ocean's environment to understand how humans and oceans affect each other.

Glossary

The first important place a term is used in the guide is noted in parentheses at the end of the definition, e.g., (Q3) refers to Question 3, How Many Phytoplankton Does It Take to Fill a Humpback Whale?

aquaculture—farming fish, shellfish, algae, or other foods from the sea (Q7)

bioluminescence—production of light by living organisms (Q5)

bloom—a sudden increase in the number of phytoplankton often following a flood of nutrients from heavy rains or a string of sunny days (Q5, Q6)

carnivore—an animal that eats other animals (Q3)

chlorophyll—the pigment used in photosynthesis to capture light energy and convert it to chemical energy (Q1)

coccolithophore—(COCK-oh-LITH-oh-for) a tiny phytoplankton; covered by many plates made of calcium carbonate (Q2)

consumer—an animal that eats producers or eats other animals that eat producers (Q3)

control—in an experiment, something with which to compare the results of the experimental test (Q1)

copepod—(KO-puh-pod) an important, abundant zooplankter; tiny (less than 1/16 in-long) crustacean (related to lobsters, crabs, krill) (Introduction, Q3)

crustacean—(cruh-STAY-shun) an animal with a tough exoskeleton, and jointed appendages and many legs (such as crabs) (Q3)

decomposer—organisms such as fungi and bacteria that break down living or dead material into chemicals that can be recycled as nutrients (Q3)

diatom—(DYE-ah-tom) phytoplankton with glass-like shell made of silica (Q2)

dinoflagellate—(dye-no-FLAH-gel-ate) phytoplankton often with two whip-like flagella; sometimes poisonous (Q2)

ecology—(ee-COHL-oh-gee) the study of the environment (Introduction, Q3)

ecosystem—the interactions of the living and non-living components of a particular area (Q3)

energy flow—the movement of energy from the sun through living organisms in a food web (Q3)

flagella—whip-like extensions of a cell; used in locomotion (Q2)

food chain—sequence of organisms in a community through which energy moves (as food); begins with the producers (photosynthesizers), then herbivores, then carnivores (Q3)

food web—complex interactions of organisms relating to who eats whom (Q3)

fossil fuel—fuels such as coal, oil and natural gas formed from fossil plants, animals, and phytoplankton (Q7)

global warming—the increase in atmospheric temperature over a long period due, in part, to an increase in the Greenhouse Effect (Q7)

Greenhouse Effect—trapping of the sun's heat energy in our atmosphere due to a layer of carbon dioxide that prevents the escape of some of the heat back into space; burning fossil fuels release carbon dioxide and may increase the Greenhouse Effect (Q7)

herbivore—an animal that eats plants (Q3)

hypothesis—a statement of prediction about the outcome of an experiment or predicted explanation for an observed event (Q1)

krill—a two-inch-long marine crustacean that grows in abundance; most important source of food for some baleen whales, particularly in the Antarctic; requires healthy phytoplankton population on which to feed (Q3)

nutrients—chemicals required for organisms to live and grow; example, phytoplankton requires nitrogen and phosphorus (Introduction, Q1)

omnivore—an animal that eats plants and other animals (Q3)

ozone—a gas present in the upper atmosphere; it helps prevent harmful ultra-violet radiation from reaching the earth's surface (Q7)

photosynthesis—process in which light energy is converted to chemical energy by plants using water, carbon dioxide; results in the production of oxygen and carbohydrates such as sugars and starches (Q1)

phytoplankter—refers to one kind of phytoplankton organism (phytoplankton refers to phytoplankton in general) (Q1)

phytoplankton—(FY-toe-plank-ton) microscopic, single-celled, drifting, photo-synthesizers not able to swim against currents (Introduction)

plankton—any aquatic organisms that drift at the mercy of the current or have very limited swimming abilities (Introduction)

producer—organisms that convert light energy to chemical energy; examples: phytoplankton, plants (Q3)

red tide—bloom of phytoplankton, usually dinoflagellates (Q6)

shellfish—aquatic molluscs (animals with very hard shells) such as clams, oysters, and mussels (Q6)

symbiosis—(sim-by-OH-sis) a close relationship between two non-related organisms in which neither is harmed, and one or both may benefit; example, clown fish and anemones or corals and zooxanthellae (Q4)

toxic—poisonous (Q6)

upwelling—movement of water (and nutrients) from deep water towards the surface; an important source of nutrients for phytoplankton growth (Q5)

zooplankton—(ZOH-plank-ton) animal plankton ranging from microscopic larval sea stars to huge jellyfish (Introduction)

zooxanthellae—(ZOH-zan-THELL-ee) phytoplankton that live in a symbiotic relationship inside the living tissue of an animal; examples, dinoflagellates living in a giant clam or in coral (Q4)

Resources for Students

A Child's Treasury of Seaside Verse. Mark Danile, (comp.). Dial. 1991.
A collection of verses about the sea from British and American writers of the 19th and early 20th centuries

Alligators to Zooplankton: A Dictionary of Water Babies. Les Kaufman and staff, New England Aquarium. Watts. 1991
Features some of the most bizarre water animal babies with information about the physical size, habitat, and the special characteristics of each animal.

Chains, Webs and Pyramids: The Flow of Energy in Nature. Laurence Pringle. Crowell. 1975.
Includes discussion of marine and salt marsh food chains.

Night of the Moonjellies. Mark Shasha. Half Moon Books, New York. 1992.
A day in the life of a child culminates in a nighttime boat trip to see bioluminescent jellies.

Pagoo. Holling Clancy Holling. Houghton Mifflin Company, Boston. 1985.
First section describes Pagoo's life as a zooplankter (hermit crab larva). Wonderful story.

Plankton: Drifting Life of the Waters. Julian May. Holiday. 1972.
Describes plant and animal plankton with emphasis on their role in food chains.

The Sea Is Calling Me. Lee Bennett Hopkins, ed. Harcourt. 1986.
A collection of poems about the sea and seashores.

Resources for Educators

A Guide to Marine Coastal Plankton and Marine Invertebrate Larvae. Deboyd Smith. Kendall/Hund. 1977.
This guide to marine plankton has a dichotomous key and is clearly illustrated with black and white plates and line drawings.

Coral Reefs. S. Walker, R. Newton, A. Ortiz. EPA 160-B-97-900a. 1997. Order online at: www.epa.gov/nceplhm/index.html English and Spanish versions of compilation of activities for middle school students.

Dictionary of Word Roots and Combining Forms. Donald J. Borror. National Press Books. Palo Alto, 1971

Discovering Density. Lawrence Hall of Science, University of California, Berkeley. 1997.
Useful resource if you choose to expand the exploration of density (and its relationship to phytoplankton).

Marine Biology, An Ecological Approach. James W. Nybakken. Harper & Row, Publishers, NY 1988

Marine Education: A Bibliography of Educational Materials. Available from any of the Sea Grant College Programs. July 1988. Marine Information Service, Sea Grant College Program, Texas A & M University, College Station, TX 77843-4115 or Maine/New Hampshire Sea Grant Program, Kingman Farm, University of New Hampshire, Durham, NH 03824 (603-749-1565).

National Science Education Standards. National Research Council. National Academy Press, Washington. 1999.
Thorough discussion and summary of the national science education standards.

Pacific Coast Pelagic Invertebrates, A Guide to the Common Gelatinous Animals.
David Wrobel and Claudia Mills. Monterey Bay Aquarium Press. 1999.
Excellent photographs of gelatinous zooplankters.

Plankton. National Geographic Video #51167. 1986.
A 12-minute video introducing the various kinds of plankton and their feeding behavior.

Ranger Rick's Naturescope: Diving Into Oceans, Activity Guide. National Wildlife Federation. 1988.
Stand-alone activities for elementary-level classes.

Reading the Environment. Mary M. Cerullo. Heinemann, Portsmouth, NH. 1997.
Useful resource for your personal library. Suggestions for using trade books to nurture a child's discovery of science.

Space Available: Aquatic Applications of Satellite Imagery. Gulf of Maine Aquarium, Portland. 1997.
Activity guide for K-12 emphasizing remote sensing, oceans, weather, and human impact.

Undersea Life. Joseph S. Levine. Stewart, Tabori and Chang, Publishers. NY. 1985.
Breathtaking photographs and excellent adult-level text. Unusual coffee-table book.

Equipment and Material Resources

Chemi-luminescence kits: Purchase from
Flinn Scientific Inc. 1-800-452-1261 Cool Light Kit or Energetic Light Kit (under $25)
Carolina Biological Supply Company 1-800-334-5551 The Cool Blue Light Kit ($7-$36)

Phytoplankton cultures: Purchase from
Carolina Biological Supply Company 1-800-334-5551
Or other biological supply company

Discovery Scopes: Purchase from
Discovery Scope, Inc., P.O.Box 607 Green Valley, AZ 85622 (602) 648-1401
A handy, hand-held, inexpensive microscope to use in classroom or field for a variety of science projects.

Websites

Bigelow Laboratory for Ocean Studies, Maine: www.bigelow.org
Subsection on "education" has phytoplankton activity and a section on "flowcam" shows recent photos of plankton; Bigelow Laboratory is a center for research on marine phytoplankton; it maintains cultures of hundreds of phytoplankton species which you may order.

Greenhouse Gases: http://www.erl.noaa.gov
The Aeronomy Laboratory of the NOAA Environmental Research laboratories studies greenhouse gases. This site might be useful for highly motivated, advanced students; look under Greenhouse Warming.

Gulf of Maine Aquarium: www.octopus.gma.org
Great resource for marine educators.

Harmful Algal Blooms: www.redtide.whoi.edu/hab/
Good source of information about blooms from Woods Hole Oceanographic Institution.

Texas Red Tides: www.tpwd.state.tx.us/fish/recreat/redtide.htm

JASON Project Homepage: http://www.jasonproject.org
Good resource for technological and interdisciplinary approach to teaching.

National Marine Educators Association: www.marine-ed.org/
A useful ocean sciences education teacher resource center bridging the gap between research and education. Join on line discussion, scuttlebutt@vims.edu, a forum for marine educators nationwide.

NOAA (National Oceanic and Atmospheric Administration):
www.websites.noaa.gov
Information about helpful NOAA websites and resources relating to oceans

National Science Foundation: www.geo.nsf.gov/
Information on world's oceans

The Nanoworld Image Gallery: www.uq.oz.au/nanoworld/images_1.html
View a variety of images from the microscopic world including phytoplankton

Office of Naval Research: www.onr.navy.mil
Useful for motivated upper level students

Satellite Images of Phytoplankton:
   http://k12science.stevens-tech.edu/curriculum/oceans/seawifs.html

Analyze satellite images of phytoplankton including a real-time data activity.
   http://seawifs.gsfc.nasa.gov/SEAWIFS/IMAGES/GALLERY.html

Images of ocean color from space include global, Atlantic coast, Australia, etc.
   www:gma.org

Gulf of Maine Aquarium satellite images; do a GMA search for satellite images (variety available).

Woods Hole Oceanographic Institution: www.whoi.edu

Organizations

Center for Marine Conservation
1725 DeSales St., NW
Washington, DC 20036
Active in protecting marine resources; good printed resources about the ocean.

Coral Reef Alliance
809 Delaware Street
Berkeley, CA 94710
http://www.coral.org/
Resource for more details about coral reefs.

FOR SEA/Marine Science Center
17771 Fjord Drive NE
Poulsbo, WA 98370
Excellent curriculum publications.

MARE: Marine Activities, Resources and Education
Lawrence Hall of Science
University of California
Berkeley, CA 94702
Excellent curriculum publications such as their *Science Teacher's Guide to the WETLANDS.*

National Marine Educators Association
PO Box 1470
Ocean Springs, MS 39566-1470

www.marine-ed.org/
A valuable organization to join; excellent regional and national meetings for teachers and other educators.

National Ocean Service/Outreach Unit
NOAA Building 4
1305 East-West Highway
Silver Springs, MD 20910
Send for list of publications; NOS includes National Marine Sanctuaries Division and National Estuarine Research Reserve Division.

National Office for Marine Biotoxins and Harmful Algal Blooms
Woods Hole Oceanographic Institution
Woods Hole, MA 02543

New England Aquarium
Teacher Resource Center
Central Wharf
Boston, MA 62110
Excellent resource library and loaner kits.

United States Environmental Protection Agency
Public Information Center
401 M Street SW
Washington, DC 20460
Send for list of publications.